LONDON MATHEMATICAL SOCIETY LECTURE
NOTE SERIES: 396

Arithmetic Differential Operators over the *p*-adic Integers

CLAIRE C. RALPH
Cornell University, USA

SANTIAGO R. SIMANCA
Université de Nantes, France

CAMBRIDGE
UNIVERSITY PRESS

CAMBRIDGE UNIVERSITY PRESS
Cambridge, New York, Melbourne, Madrid, Cape Town,
Singapore, São Paulo, Delhi, Tokyo, Mexico City

Cambridge University Press
The Edinburgh Building, Cambridge CB2 8RU, UK

Published in the United States of America by Cambridge University Press, New York

www.cambridge.org
Information on this title: www.cambridge.org/9781107674141

First published 2012

Printed in the United Kingdom at the University Press, Cambridge

A catalogue record for this publication is available from the British Library

ISBN 978-1-107-67414-1 Paperback

l o o ³ ᵬ ᒲ Ƨ Ƨ ᑫ ₆ Ƨ ᔿ Ƨ ᔿ Ƨ ᔿ Ƨ ᔿ

Contents

1

Introduction

Our purpose in this monograph is to provide a concise and complete introduction to the study of arithmetic differential operators over the p-adic integers $\mathbb{Z}_p$. These are the analogues of the usual differential operators over say, the ring $\mathbb{C}[x]$, but where the role of the variable x is replaced by a prime p, and the roles of a function $f(x)$ and its derivative df/dx are now played by an integer $a \in \mathbb{Z}$ and its Fermat quotient $\delta_p a = (a - a^p)/p$.

In making our presentation of these type of operators, we find no better way than discussing the p-adic numbers in detail also, and some of the classical differential analysis on the field of p-adic numbers, emphasizing the aspects that give rise to the philosophy behind the arithmetic differential operators. The reader is urged to contrast these ideas at will, while keeping in mind that our study is neither exhaustive nor intended to be so, and most of the time we shall content ourselves by explaining the *differential* aspect of an arithmetic operator by way of analogy, rather than appealing to the language of jet spaces. But even then, the importance of these operators will be justified by their significant appearance in number theoretic considerations. One of our goals will be to illustrate how different these operators are when the ground field where they are defined is rather coarse, as are the p-adic integers $\mathbb{Z}_p$ that we use.

In order to put our work in proper perspective, it is convenient to introduce some basic facts first, and recall a bit of history. Given a prime p, we may define the p-adic norm $\| \ \|_p$ over the field of rational numbers $\mathbb{Q}$. The completion of the rationals in the metric that this norm induces is the field $\mathbb{Q}_p$ of p-adic numbers, and this field carries a non-Archimedean p-adic norm extending the original p-adic norm on $\mathbb{Q}$. This is the description of $\mathbb{Q}_p$ as given by K. Hensel circa 1897 (see, for instance,

1

[28]). Two decades later, A. Ostrovski [39] proved that any nontrivial norm on $\mathbb{Q}$ is equivalent to either the Euclidean norm or to a p-adic norm for some prime p. In this way, there arose the philosophical principle that treats the real numbers and all of the p-adic numbers on equal footing.

In the twentieth century, the p-adic numbers had a rich history. We briefly mention some major results.

The idea that studying a question about the field $\mathbb{Q}$ can be answered by putting together the answers to the same question over the fields $\mathbb{R}$ and $\mathbb{Q}_p$ for all ps was born with the Hasse–Minkowski's theorem. This states that a quadratic form over $\mathbb{Q}$ has a nontrivial zero in $\mathbb{Q}^n$ if, and only if, it has a nontrivial zero in $\mathbb{R}^n$ and a nontrivial zero in $\mathbb{Q}_p^n$ for each prime p. This theorem was proven by Hasse in his thesis around 1921 [27], the problem having been proposed to him by Hensel who had proven the $n = 2$ case a few years earlier. Such a principle fails for cubics.

The development above came after several interesting results that preceded the introduction of the p-adic numbers. The local-to-global principle embodied in the Hasse–Minkowski theorem had a precedent in Riemannian geometric, since as recently as 1855, Bonnet had proved that if the curvature of a compact surface was bounded below by a positive constant, then its diameter was bounded above by a quantity depending only on the said constant. Strictly on the arithmetic side of things, in the seventeenth century J. Bernoulli defined the Bernoulli numbers B_k, the coefficients in the expansion $e^t/(e^t - 1) = \sum_k B_k t^k/k!$, used them to compute closed-form expressions for the sums $\sum_{j=0}^m j^n$, and developed several identities that these numbers satisfy. A century later, the Bernoulli numbers were used by Euler to show heuristically that if ζ is the Riemman zeta function, then $\zeta(1 - k) = \sum_{n=1}^\infty 1/n^{1-k} = -B_k/k$ for any integer $k \geq 2$. In the mid nineteenth century, Riemman proved that $\zeta(s) = \sum_{n=1}^\infty 1/n^s$ is a meromorphic function on the complex plane $\mathbb{C}$, giving Euler's argument complete sense. Further, he used the Gamma function to define $\Lambda(s) = \pi^{-\frac{s}{2}} \Gamma\left(\frac{s}{2}\right) \zeta(s)$ and proved the functional equation $\Lambda(s) = \Lambda(1 - s)$. The intimate relationship between the Bernoulli numbers and the values of $\zeta(s)$ at negative integers led to the idea that these numbers have profound arithmetical properties, a fact discovered by Kummer in his work on Fermat's last theorem circa 1847. The ideal class group of $\mathbb{Q}(\zeta_N)$, ζ_N a primitive N-th root of unity, is the quotient of the fractional ideals of $\mathbb{Q}(\zeta_N)$ by the set of principal ideals, and it turns out to be a group of finite order h_N with respect to ideal multiplication. A prime p is said to be regular if $p \nmid h_p$, and irregular otherwise. Kummer proved that p is regular if, and only if, p does not divide the numerator of $B_2, B_3, \ldots, B_{p-3}$ and that Fermat last theorem holds for all regular primes. He also proved that, if $m \equiv n \not\equiv 0 \bmod p - 1$, then $B_m/m \equiv B_n/n \bmod p$, the congruences that are nowadays named after him. They led to the proof that that there are infinitely many irregular primes. Since heuristically it can be proven that there is a large percentile of regular primes, Kummer's ideas had remarkable implications in the study of Fermat's last theorem. Thus, algebraic number theory and the theory of L-functions were born and replaced the elementary methods used before him in the analysis of this problem.

C. Chevalley defined the *adèle* ring and *idèle* group [20], and used them to reformulate class field theory [21] around 1932. For convenience, if we denote by $\| \ \|_\infty$ the Euclidean norm in $\mathbb{R}$, which we think of as $\mathbb{Q}_\infty$, the field of p-adic numbers corresponding to $p = \infty$, we take the Cartesian product $\mathbb{Q}_\infty \times \prod_p \mathbb{Q}_p$, and define the adèle ring $\mathbb{A}_\mathbb{Q}$ to be

$$\mathbb{A}_\mathbb{Q} = \left\{ (a_\infty, a_2, a_3, a_5, \ldots) \in \mathbb{Q}_\infty \times \prod_p \mathbb{Q}_p : \|a_p\|_p \leq 1 \text{ for almost all } ps \right\}.$$

Its ring structure is obtained by defining addition and multiplication component-wise; it contains an isomorphic image of $\mathbb{Q}$ via the mapping

$$\mathbb{Q} \ni q \overset{a_\mathbb{Q}}{\to} (q, q, \ldots) \in \mathbb{A}_\mathbb{Q}.$$

For $a \in \mathbb{Q}_\infty \times \prod_p \mathbb{Q}_p$, we define $\|a\|_p = \|a_p\|_p$. Then $a \in \mathbb{A}_\mathbb{Q}$ if, and only if, $\|a\|_p \leq 1$ for all but finitely many ps. The subset $I_\mathbb{Q}$ of $\mathbb{A}_\mathbb{Q}$ consisting of all as such $\|a\|_p \neq 0$ for all ps, and $\|a\|_p = 1$ for all but finitely many of them, is the idèle multiplicative group. It contains an isomorphic image of $\mathbb{Q}^\times$ by restriction of the mapping $a_\mathbb{Q}$ above. If F is an extension of $\mathbb{Q}$, the norms on $\mathbb{Q}$ can be extended to norms on F, and we naturally define I_F also. There is a norm homomorphism $I_F \to I_\mathbb{Q}$, and its image $N(I_F/I_\mathbb{Q})$ is a group. The Galois group of $F/\mathbb{Q}$ is naturally isomorphic to $I_\mathbb{Q}/\mathbb{Q}^\times N(I_F/I_\mathbb{Q})$. Chevalley proved this fact using the local theory, avoiding the use of tools from analytic number theory. He generalized it also for number fields, fields that are extensions of $\mathbb{Q}$ of infinite degree.

In his thesis, J. Tate used real harmonic analysis on the adèles to prove functional equations for the Riemann zeta function. T. Kubota and H.W. Leopoldt [32] introduced a p-adic version of the Riemann zeta function, and used it to interpret Kummer's congruences for Bernoulli numbers mentioned above, which date back to 1851. Y.I. Manin and B. Mazur [38] interpreted the result of Kubota and Leopoldt in terms of a p-adic Mellin transform, and found p-adic interpretations of L-

functions of elliptic curves. The p-adic integers $\mathbb{Z}_p$ were known to appear as Galois groups of some infinite cyclotomic extensions. K. Iwasawa considered the completed group algebras of these Galois groups, which act on class groups and make them modules over the completed groups. These modules have some invariants. Iwasawa conjectured that these invariants could be read off from classical Dirichlet L-functions after a p-adic interpolation, using the p-adic Mellin transform. This conjecture was proved by B. Mazur and A. Wiles [33]. Triggered by the work of Tate, B. Dwork studied p-adic differential equations, and gave a p-adic proof of the rationality of Weil's zeta function [23], taking then a major step in the settling of all of the Weil conjectures about this function [48], work that was completed by P. Deligne [22]. J.-P. Serre and N. Katz studied several other p-adic functions of arithmetic interest, and A. Grothendieck studied p-adic cohomology and crystalline cohomology.

The list of problems in the field is outstanding, and the list of contributors to their understanding and resolution is important. We have not come even close to exhausting either one. But we can now retake the main theme of our work in this introduction with a better perspective in mind.

In the course of modern mathematical history, analogies between functions and numbers have played an important role in the development of number theory. The fundamental theorems of algebra and arithmetic can be seen as counterparts to each other, with the integers -1, 0 and 1 playing the role of the constant polynomials in $\mathbb{C}[x]$. This point of view is once again motivational to the philosophy of arithmetic differential operators, the idea at the level of the integers $\mathbb{Z}$ being to find an appropriate substitute $\delta_p : \mathbb{Z} \to \mathbb{Z}$ for the derivative operator

$$\partial_x = \frac{d}{dx} : \mathbb{C}[x] \to \mathbb{C}[x].$$

Indeed, given a "number" x, lets us think of it as a "function," and consider the expression $x - x^p$, one that makes frequent appearances in number theoretic considerations. For $\mathbb{F}_p$, the finite field of p elements, the identity $x - x^p = 0$ holds for all elements. In the more general situation, we can restrict our attention to numbers such that $x - x^p \equiv 0 \mod p$. We think of x as a function of p, and interpret the difference $x - x^p$ as the variation of x as its argument changes to p. We then use the Fermat quotient

$$\delta_p x = \frac{x - x^p}{p} \tag{$\dagger$}$$

to define the notion of arithmetic derivative of x in the direction of p. This is the notion that we shall be studying here, most of the time restricting our attention to xs that are taken from the ring of p-adic integers $\mathbb{Z}_p$. At this point, though, this quotient is just a heuristic statement.

The theory of arithmetic differential operators that ensues from the idea outlined above was proposed by A. Buium [6, 8], with δ_p serving in the role of the arithmetic analogue of the operator ∂_x on the polynomial ring $\mathbb{C}[x]$. At the purely arithmetic level, it serves also as a substitute for Dwork's operator

$$\frac{d}{dx} : \overline{\mathbb{F}}_p[x] \to \overline{\mathbb{F}}_p[x]$$

in his theory of p-adic differential equations over the differential field $\overline{\mathbb{F}}_p[x]$, $\overline{\mathbb{F}}_p$ the algebraic closure of the field $\mathbb{F}_p$ with p-elements. In Dwork's theory [25], the xs are still being viewed as an "argument to the functions" rather than as functions themselves. But the arithmetic differential operator δ_p exhibits an additional fundamental difference with the Dwork's operator that is worth pointing out now: δ_p is highly nonlinear, with additivity holding only modulo a lower order term measured by a polynomial with integer coefficients, and a Leibniz rule that holds but only highly intertwined with the p-th power homomorphism, and modulo terms that are p-adically smaller.

In fact, more can be said at this point. If we were to develop a differential theory with operators of the type

$$u \mapsto P\left(u, \frac{du}{dx}, \ldots, \frac{d^r u}{dx^r} \right)$$

where $P(x_0, \ldots, x_r)$ is a polynomial function, we would obtain the Ritt–Kolchin theory of "ordinary differential equations" with respect to $\frac{d}{dx}$, cf. [41, 30, 19]. This would lead to the notion of the $\frac{d}{dx}$-*character* of an algebraic group, which should be viewed as the analogue of a linear ordinary differential operator on an algebraic group (cf. to the Kolchin logarithmic derivative of algebraic groups defined over $\widehat{\mathbb{Z}}_p^{ur}$, [30, 19], and the Manin maps of Abelian varieties defined over $\widehat{\mathbb{Z}}_p^{ur}[[q]]$ [38, 12], $\widehat{\mathbb{Z}}_p^{ur}$ the unramified completion of the ring of p-adic integers).

If instead we were to develop a theory with operators that are the p-adic limits of $P(u, \delta_p u, \ldots, \delta_p^r u)$, $P(x_0, \ldots, x_r)$ a polynomial, we would then obtain the arithmetic analogue of the ordinary differential equations of Buium, as found in [8, 9, 6]. In particular, we would then arrive at the notion of a δ_p-character of a group scheme, the arithmetic "version" of a linear ordinary differential equation on a group scheme.

In this monograph, we apply and study Buium's idea over the rather coarse ring of p-adic integers $\mathbb{Z}_p$. We think of the elements in this ring as functions over a space of dimension zero that vary *infinitesimally* according to the heuristic equation (†) at the prime p. The ensuing notion of derivative is the one alluded to in the title, and on which we shall elaborate extensively in what follows. We will pause at some point to define these arithmetic operators with the generality given in Buium's work. This will benefit the interested reader while allowing us to contrast the behaviour of these operators when defined over $\mathbb{Z}_p$ or $\widehat{\mathbb{Z}}_p^{ur}$. Ultimately, it is the fact that we can cast these operators as global functions on a suitable arithmetic jet space, for any smooth scheme of finite type, which allows for their interpretation as differential operators of sorts, the way the usual differential operators on a manifold are sections of its jet bundles.

Given such a notion of arithmetic derivative, we then may define in the obvious manner an arithmetic differential operator of order n, where n is an arbitrary positive integer. Over the ring of p-adic integers $\mathbb{Z}_p$, we have also the classical notion of an analytic function. We shall show that all arithmetic differential operators turn out to be analytic functions. Quite remarkably in fact, characteristic functions of p-adic discs are shown to be equal to arithmetic operators of an order that depends upon the radius of the disc, generalizing a result that we first prove via an explicit construction, namely that the characteristic function of a disc of radius $1/2$ over the 2-adic integers is an arithmetic differential operator of order one. The extended result for a general prime is a bit surprising, point upon which we will elaborate in due course.

We organize our work as follows: in Chapter 2, we summarize the construction of the p-adic numbers and the p-adic integers, describe its topology as a metric space, its analytic and algebraic properties, and study the $(p-1)$-roots of unity in it. In Chapter 3 we study some results from classical analysis on $\mathbb{Q}_p$, including Mahler's theorem that establishes a bijection between the sets of restricted sequences and that of continuous functions on $\mathbb{Z}_p$, we present basic properties of the Artin–Hasse function, and study the analytic completion of the algebraic closure of $\mathbb{Q}_p$, the p-adic alter ego of the complex numbers that result when we complete $\mathbb{Q}$ in the Euclidean metric instead. In Chapter 4 we introduce the set of analytic functions as a required preliminary to our discussion later on. The arithmetic differential operators make their

LONDON MATHEMATICAL SOCIETY LECTURE NOTE SERIES

Managing Editor: Professor M. Reid, Mathematics Institute,
University of Warwick, Coventry CV4 7AL, United Kingdom

The titles below are available from booksellers, or from Cambridge University Press
at http://www.cambridge.org/mathematics

first appearance in Chapter 5, where we tie them to their associated homomorphisms. This in turn allows us to prove that equation (†) defines the only arithmetic differential operator over $\mathbb{Z}_p$ since this ring carries just one automorphism. Using it as a building block, we define an arithmetic differential operator of any order. We discuss also the basic rings that must be used in the theory when we have multiple primes, essentially to indicate the additional difficulties that arise then. In Chapter 6 we pause to define arithmetic operators in general, developing succinctly the theory of arithmetic jet spaces of Buium. In order to make things easier for analysts not accustomed to algebraic concepts, we present a list of the concepts from commutative algebra and schemes that are needed in the development of the general theory. In the case of group schemes, we discuss the characters that have been alluded to earlier, the analogs in the theory of the *linear* differential operators. And we outline the theory for multiple primes also, in a succinct manner. In Chapter 7 we prove that all arithmetic differential operators over $\mathbb{Z}_p$ are analytic functions. In Chapter 8 we study characteristic functions of p-adic discs from the point of view of the theory of arithmetic differential operators, and prove that they are indeed differential operators of an order depending upon the radii of the discs. The prime $p = 2$ manifests itself in a rather special manner here, as we are able to prove by way of a direct argument that the characteristic function of a discs of radius $1/2$ over the 2-adic integers is an arithmetic differential operator of order one. This work is carried out in *standard coordinates*, and leads to some formal power series representations of the characteristic functions when the prime in question is singular, a concept that we define then. In Chapter 9 we work with *harmonic coordinates*, and improve significantly upon the result in the previous chapter, showing that all analytic functions on $\mathbb{Z}_p$ are arithmetic differential operators, with the order being equal to the level of analiticity. This last concept had made its first appearance earlier, in the context of Chapter 4. Finally, in Chapter 10, we exhibit some fundamental differences in the behavior of arithmetic differential operators that manifest when we work over the ring $\widehat{\mathbb{Z}}_p^{ur}$ instead of $\mathbb{Z}_p$. In particular, we indicate how to show that as soon as we adjoin one unramified root of unity to $\mathbb{Z}_p$, the counterpart of the theorem above on the characteristic function of discs no longer holds.

2

The p-adic numbers $\mathbb{Q}_p$

The field $\mathbb{Q}_p$ of p-adic numbers is the completion of the field $\mathbb{Q}$ of rational numbers with respect to the p-adic norm. In this chapter, we explain their construction from various points of view, all, of course, equivalent to each other.

Let $p \in \mathbb{Z}$ be a prime that we fix hereafter. For $a \in \mathbb{Z}$, we let $\mathrm{ord}_p\, a$ be the exponent of p in the prime factorization of a, that is to say, the integer l such that $a = p^l r$, where $r \in \mathbb{Z}$ is not divisible by p. This notion is extended to a rational number $q = a/b$ by setting $\mathrm{ord}_p\, q = \mathrm{ord}_p\, a - \mathrm{ord}_p\, b$, and the resulting function is multiplicative, that is to say, it has the property that $\mathrm{ord}_p\, q_1 q_2 = \mathrm{ord}_p\, q_1 + \mathrm{ord}_p\, q_2$. We then define the p-adic norm function on $\mathbb{Q}$ by

$$\|q\|_p = \frac{1}{p^{\mathrm{ord}_p\, q}}. \tag{2.1}$$

We shall denote by d_p the metric on $\mathbb{Q}$ that this norm induces.

In the resulting norm on $\mathbb{Q}$, a rational q has $\|q\|_p \leq 1$ if, and only if, the denominator b of its reduced rational form a/b is not divisible by p. Integers are closer to each other in the metric d_p on $\mathbb{Q}$ the higher the power of p that divides their difference. So, for instance,

$$d_5(2,1) = \|1 - 2\|_5 = 1, \text{ while } d_5(2,127) = \|2 - 127\|_5 = \frac{1}{5^3}.$$

The p-adic norm satisfies a condition stronger than the triangle inequality. Indeed, if $q = a/b$ and $r = c/d$, since the biggest power of p that divides $ad + bc$ is at least the minimum of the biggest power dividing

8

ad and the biggest power dividing bc, we have that

$$
\begin{aligned}
\mathrm{ord}_p\,(q+r) \;&=\; \mathrm{ord}_p\left(\frac{ad+bc}{bd}\right)\\
&\geq\; \min\left\{\mathrm{ord}_p\,ad, \mathrm{ord}_p\,bc\right\} - \mathrm{ord}_p\,b - \mathrm{ord}_p\,d\\
&=\; \min\left\{\mathrm{ord}_p\,a + \mathrm{ord}_p\,d, \mathrm{ord}_p\,b + \mathrm{ord}_p\,c\right\} - \mathrm{ord}_p\,b - \mathrm{ord}_p\,d\\
&=\; \min\left\{\mathrm{ord}_p\,a - \mathrm{ord}_p\,b, \mathrm{ord}_p\,c - \mathrm{ord}_p\,d\right\}\\
&=\; \min\left\{\mathrm{ord}_p\,q, \mathrm{ord}_p\,r\right\},
\end{aligned}
$$

and therefore,

$$
\|q+r\|_p = \frac{1}{p^{\mathrm{ord}_p\,(q+r)}} \leq \max\left\{p^{-\mathrm{ord}_p\,q}, p^{-\mathrm{ord}_p\,r}\right\} = \max\left\{\|q\|_p, \|r\|_p\right\}.
$$
(2.2)

The triangle inequality now follows readily. This stronger inequality (2.2), referred to as the "non-Archimedean property" of $\|\ \|_p$, produces some geometric results that contrast a bit with those from our more traditional point of view in Euclidean geometry. Triangles, for instance, are all isosceles.

For let us assume that we have a "triangle with vertices at 0, q and r," respectively. We then know that $\|q-r\|_p \leq \max\left\{\|q\|_p, \|r\|_p\right\}$. If $\|q\|_p < \|r\|_p$, the non-Archimedean property of the norm implies that

$$
\|q-r\|_p \leq \|r\|_p\,.
$$

Since $\|r\|_p = \|q-(q-r)\|_p \leq \max\left\{\|q\|_p, \|q-r\|_p\right\}$, it follows that

$$
\|r\|_p \leq \|q-r\|_p
$$

also. Thus, $\|r\|_p = \|q-r\|_p$ and so, in the geometry generated by $\|\ \|_p$, all triangles are isosceles, with the two largest sides equal to each other in length. Sometimes we shall refer to this as the "isosceles triangle property" of $\|\ \|_p$.

Now we describe briefly the general process that defines $\mathbb{Q}_p$ as the metric completion of $\mathbb{Q}$ in the distance defined by the p-adic norm. This yields $\mathbb{Q}_p$ as the unique complete field, up to isometric isomorphism, that contains a $\|\ \|_p$-isometric dense copy of the field $\mathbb{Q}$.

We say that a sequence $\{q_n\}$ of rational numbers is Cauchy with respect to the norm $\|\ \|_p$, if for any real number $\varepsilon > 0$ there exists N such that $\|q_n - q_m\|_p < \varepsilon$ for all $n, m \geq N$. We say that the sequence $\{q_n\}$ is *null* if for any $\varepsilon > 0$ there exists N such that $\|q_n\|_p < \varepsilon$ for all $n \geq N$.

Given rational Cauchy sequences $\{q_n\}$ and $\{r_n\}$, we define their addition and multiplication by

$$
\{q_n\} + \{r_n\} = \{q_n + r_n\}, \quad \{q_n\}\{r_n\} = \{q_n r_n\}.
$$

Let R be the set of all rational Cauchy sequences, and let M be the subset of all null sequences. The operations above provide R with a ring structure, and M becomes an ideal in R. In fact, M is a maximal ideal. For if $\{q_n\} \in R$ is not null, there exists an $\varepsilon > 0$ and an integer N such that $\|q_n\|_p > \varepsilon$ for any $n > N$, and we may set

$$r_n = \begin{cases} 0 & \text{for} \quad n \le N\,, \\ \dfrac{1}{q_n} & \text{for} \quad n > N\,. \end{cases}$$

This is a Cauchy sequence also, and we have

$$\{q_n\}\{r_n\} = \{0, \ldots, 0, 1, 1, \ldots\} = \{1, 1, \ldots\} - \{1, \ldots, 1, 0, 0, \ldots\}\,.$$

If I were an ideal with $M \subset I$ properly, then I would contain a non-null sequence $\{q_n\}$. Since the sequence $\{1, \ldots, 1, 0, 0, \ldots\}$ is null, the argument above with r_n would imply that the constant sequence $\{1, 1, \ldots\}$ must be contained in I, and so I would have to be equal to R. Thus, M is a maximal ideal. The quotient field R/M is, by definition, the field of p-adic numbers $\mathbb{Q}_p$.

An additional detail exhibits a fundamental difference between this construction of $\mathbb{Q}_p$ and the analogous construction of $\mathbb{R}$ as the completion of $\mathbb{Q}$ in the Euclidean norm. Given a p-adic number $\alpha = \{q_n\} + M$, if the Cauchy sequence $\{q_n\}$ is null, we set $\|\alpha\|_p = 0$. Otherwise, there exist a positive real number ε and an integer N such that $\|q_n\|_p > \varepsilon$ for any $n > N$. We may choose N sufficiently large so that $\|q_n - q_m\|_p < \varepsilon$ for $n, m > N$ also. Then, by the isosceles triangle property, we have that $\|q_n\|_p$ is constant for all $n > N$, and so we may define

$$\|\alpha\|_p = \lim_{n \to \infty} \|q_n\|_p\,,$$

extending in this manner the p-adic norm on $\mathbb{Q}$ to a p-adic norm on $\mathbb{Q}_p$. Using this extension, we may also extend the notion of p-adic order.

In the construction of $\mathbb{R}$ as the completion of $\mathbb{Q}$ in the Euclidean norm, this said norm admits an extension to a norm on $\mathbb{R}$ as well. However, in spite of the fact that their constructions derive from exactly the same procedure, this and the p-adic norm above exhibit a substantial difference. For unlike the case of the Euclidean norm on $\mathbb{R}$, the extended $\| \; \|_p$-norm on $\mathbb{Q}_p$ still ranges over the set $\{0\} \cup \{p^n\}_{n \in \mathbb{Z}}$, the same range this norm function has over $\mathbb{Q}$, whereas, for $\mathbb{R}$, the range of the Euclidean norm is $\mathbb{R}$ itself.

We have the mapping

$$
\begin{aligned}
\mathbb{Q} &\xrightarrow{\iota} \mathbb{Q}_p = R/M \\
q &\mapsto \{q\} + M\,,
\end{aligned}
$$

where $\{q\}$ stands for the constant sequence all of whose terms are equal to q. It defines an isometric isomorphism onto its image, which by abuse of notation we denote by $\mathbb{Q}$ also. It is dense in $\mathbb{Q}_p$. For if $\alpha = \{q_n\} + M \in \mathbb{Q}_p$, and $\varepsilon > 0$ is arbitrary, there exists some N such that $\|q_n - q_m\|_p < \varepsilon$ for $n, m > N$. Then for any $n > N$, the constant rational sequence $\beta = \{q_n, q_n, q_n, \ldots\}$ is such that $\|\alpha - \beta\|_p = \lim_{m \to \infty} \|q_m - q_n\|_p \leq \varepsilon$.

That $\mathbb{Q}_p$ is complete follows by construction. If α_n is a Cauchy sequence in $\mathbb{Q}_p$, we may find a sequence $q_1, q_2, \ldots$ in $\mathbb{Q}$ such that

$$
\|\alpha_n - q_n\|_p < \frac{1}{n}\,, \quad n = 1, 2, \ldots.
$$

Since

$$
\|q_n - q_m\|_p \leq \|q_n - \alpha_n\|_p + \|\alpha_n - \alpha_m\|_p + \|\alpha_m - q_m\|_p\,,
$$

the said rational sequence is Cauchy also, and its limit in $\mathbb{Q}_p$ exists. We set $\lim_{n \to \infty} q_n = \alpha$, so $\alpha = \{q_n\} + M$. Since

$$
\|\alpha - \alpha_n\|_p \leq \|\alpha - q_n\|_p + \|q_n - \alpha_n\|_p\,,
$$

we see that the original sequence α_n has limit α in $\mathbb{Q}_p$.

2.1 A pragmatic realization of $\mathbb{Q}_p$

It is convenient to have a pragmatic description of the elements of $\mathbb{Q}_p$ that allows us to carry calculations with some ease. We show one of these next, describing a p-adic number in a way analogous to the description of a real number by, say, its decimal expansion.

Let $\alpha \in \mathbb{Q}_p$. Since its p-adic norm is some integer power of p, we may write α as

$$
\alpha = p^n u\,, \quad \text{where} \quad \|\alpha\|_p = \|p\|_p^n\,, \quad \|u\|_p = 1\,. \tag{2.3}
$$

The unit u is the limit of some rational sequence $\{r_j\}$. Let N be an integer such that $\|u - r_j\|_p < 1$ for all $j \geq N$. By the isosceles triangle property, we have that $\|u\|_p = \|r_j\|_p = 1$ for these js. Thus, r_j is a rational unit, and

$$
u + M = r_j + M\,.
$$

We set $q_n = r_N$.

Since q_n is a rational unit, $q_n = a_n/b_n$, where a_n and b_n are relatively prime to p. In particular, there exist integers v, w such that

$$vb_n + wp = 1 \,,$$

and v is the inverse of b_n mod p. Therefore,

$$q_n - a_n v = \frac{a_n}{b_n} - a_n v = \frac{a_n(1 - vb_n)}{b_n} \,,$$

and the right side has p-adic norm less than 1.

We denote by $\mathfrak{P}$ the set of p-adic numbers of p-adic norm less than 1. If we set $c_n = a_n v \in \mathbb{Z}$, we then have that

$$u + \mathfrak{P} = q_n + \mathfrak{P} = c_n + \mathfrak{P}$$

and

$$\|u - c_n\|_p < 1 \,, \quad \|\alpha - c_n p^n\|_p < \|p^n\|_p \,.$$

Thus,

$$\alpha = u p^n = c_n p^n + (u - c_n)p^n = c_n p^n + \alpha_1 \,,$$

where $\alpha_1 = (u - c_n)p^n$ is a p-adic number such that $\|\alpha_1\|_p < \|p\|_p^n$. We must then have $\|\alpha_1\|_p = \|p\|_p^m$ for some $m > n$, and we can iterate the procedure started in (2.3) with the role of α now played by α_1. Since $\|p\|_p^{n+j} \to 0$ as $j \to \infty$, we conclude the following.

Proposition 2.1 *Any p-adic number α can be written in the form*

$$\alpha = \sum_{j=n}^{\infty} c_j p^j \,, \tag{2.4}$$

where each c_j is an integer, and $\|\alpha\|_p = \|p\|_p^n$. The representation is unique if each coefficient c_j is chosen in the range $0 \le c_j \le p - 1$.

The expansion

$$\alpha = \frac{c_{-n}}{p^n} + \frac{c_{-n+1}}{p^{n-1}} + \cdots + c_0 + c_1 p + c_2 p^2 + \cdots$$

is usually abbreviated as

$$\alpha = c_{-n} c_{-n+1} \ldots c_0 . c_1 c_2 \ldots \,.$$

Example 2.2   We find the expansion of $1/5$ in $\mathbb{Q}_3$. This rational has 3-adic norm 1, and the only integer c_0 in the range $0 \le c_0 < 3$ such that the 3-adic norm of $1/5 - c_0$ is less than 1 is $c_0 = 2$. We have

$$\frac{1}{5} = 2 + \left(\frac{1}{5} - 2\right) = 2 - \frac{9}{5} = 2 - \frac{1}{5}3^2 \,. \tag{2.5}$$

Hence, the coefficient c_1 is zero, $c_2 = 1$ because

$$-\frac{1}{5} = 1 + \left(-\frac{1}{5} - 1\right) = 1 - \frac{2}{5}3 \,,$$

and we have

$$\frac{1}{5} = 2 + 1 \cdot 3^2 - \frac{2}{5}3^3 \,.$$

Now we see that

$$-\frac{2}{5} = 2 - \frac{4}{5}3 = 2 + 1 \cdot 3 - \frac{1}{5}3^3 \,,$$

and therefore

$$\frac{1}{5} = 2 + 1 \cdot 3^2 + 2 \cdot 3^3 + 1 \cdot 3^4 - \frac{1}{5}3^6 \,.$$

Thus, $c_3 = 2$, $c_4 = 1$, $c_5 = 0$, and comparing with (2.5), we conclude that $c_6 = c_2$, and that the 3-adic expansion of $1/5$ must be periodic, with period 0121. In other words, we have

$$\frac{1}{5} = 2.01210121\ldots$$

in $\mathbb{Q}_3$. $\square$

Using the fact that

$$1 + p^k + p^{2k} + \cdots + p^{(l-1)k} = \frac{1 - p^{lk}}{1 - p^k}$$

converges p-adically to $\frac{1}{1-p^k}$, and ideas similar to the one above, we can show that $\alpha \in \mathbb{Q}_p$ is an element of $\mathbb{Q}$ if, and only if, its canonical expansion (2.4) is periodic.

2.2 The p-adic integers $\mathbb{Z}_p$ and their field of fractions

The set $\mathbb{Z}_p$ of p-adic integers is defined by

$$\mathbb{Z}_p = \{\alpha \in \mathbb{Q}_p \,:\, \|\alpha\|_p \le 1\} \,,$$

the set of p-adic numbers whose canonical expansion (2.4) begins at $n = 0$ or above. By the non-Archimedean property (2.2) of the p-adic norm, $\mathbb{Z}_p$ is a subring of $\mathbb{Q}_p$. From this point on, the elements of $\mathbb{Z}$ will be referred to as rational integers, reserving the term integer for the elements of $\mathbb{Z}_p$ instead.

We shall say that two p-adic numbers α, β are congruent mod p^n if $\|\alpha - \beta\|_p \leq p^{-n}$, and write $\alpha \equiv \beta$ mod p^n. This is equivalent to saying that $(\alpha - \beta)/p^n \in \mathbb{Z}_p$, that is to say, the expansion of $\alpha - \beta$ contains a nonzero coefficient no sooner than in the p^n-th position. Clearly, this notion extends the congruence mod p^n over $\mathbb{Z}$.

We denote by $p^n \mathbb{Z}_p$ the ideal generated by p^n in $\mathbb{Z}_p$. The congruence $\alpha \equiv \beta$ mod p^n simply says that $\alpha - \beta \in p^n \mathbb{Z}_p$. The set $\mathfrak{P}$ of p-adic numbers of norm less than 1 used in the previous section is just $p\mathbb{Z}_p$, a maximal ideal in $\mathbb{Z}_p$.

If $\alpha = \sum_{j=0}^{\infty} c_j p^j$ is an element of $\mathbb{Z}_p$, we let $a_n = \sum_{j=0}^{n-1} c_j p^j$. This defines a sequence of rational integers, and

$$\alpha - a_n = p^n \sum_{j=n}^{\infty} c_j p^{j-n} = p^n \beta$$

where $\beta \in \mathbb{Z}_p$. The said sequence $\{a_n\} \subset \mathbb{Z}$ satisfies the congruences

$$\alpha \equiv a_n \mod p^n, \quad n = 1, 2, \ldots \tag{2.6}$$

and

$$a_{n+1} \equiv a_n \mod p^n,$$

for each $n \geq 1$. Any p-adic integer satisfies an infinite set of congruences of this form, and these congruences characterize the p-adic integer in question.

Example 2.3 Let us consider the sequence of rational integers given by

$$a_n = \sum_{j=0}^{n-1} 3 \cdot 10^j.$$

We have that $a_{n+1} - a_n = 3 \cdot 10^n$, and so $a_{n+1} \equiv a_n$ mod 5^n. Thus, the said sequence yields a 5-adic number α as $n \to \infty$. This limit is in fact the number $-1/3$. For we have that $3a_n = 10^n - 1$, and so $\|3a_n + 1\|_5 \to 0$.

Of course, we could have used the prime 2 instead, and show that the said sequence converges to $-1/3$ in $\mathbb{Q}_2$ also. $\qquad\square$

The discussion above can be reversed [42]. For if $A_n = \mathbb{Z}/p^n\mathbb{Z}$ is the ring of classes of integers mod p^n, we obtain an obvious and natural surjective homomorphism

$$\varphi_n : A_n \mapsto A_{n-1} \,,$$

with kernel $p^{n-1}A_n$, and the sequence

$$\cdots \to A_n \to A_{n-1} \to \cdots \to A_2 \to A_1 \tag{2.7}$$

forms a projective system indexed by the whole numbers. Then we may define $\mathbb{Z}_p$ to be the projective limit of the system (A_n, φ_n) above, $\mathbb{Z}_p = \varprojlim(A_n, \varphi_n)$, where by definition, an element α of this limit is a sequence $\{a_n\}$ with $a_n \in A_n$ such that $\varphi_n(a_n) = a_{n-1}$. Addition and multiplication are defined component-wise. If each A_n is given the discrete topology, $\mathbb{Z}_p$ is a subring of the space $\prod A_n$ endowed with the product topology, which is compact. Thus, $\mathbb{Z}_p$ inherits a topology that makes it into a compact space, being a closed subspace of the compact ambient space where it is embedded.

If $\varepsilon_n : \mathbb{Z}_p \mapsto A_n$ is the n-th component function, we have the short exact sequence

$$0 \to \mathbb{Z}_p \xrightarrow{p^n} \mathbb{Z}_p \xrightarrow{\varepsilon_n} A_n \to 0 \,,$$

which permits to realize the identification of $\mathbb{Z}_p/p^n\mathbb{Z}_p$ with $A_n = \mathbb{Z}/p^n\mathbb{Z}$. The congruences (2.6) exhibit explicitly this identification.

The field $\mathbb{Q}_p$ is the field of fractions of $\mathbb{Z}_p$, $\mathbb{Q}_p = \mathbb{Z}_p[p^{-1}]$.

2.3 The topology of $\mathbb{Q}_p$

As a metric space, the topology of $\mathbb{Q}_p$ is given by the basis of open sets consisting of "discs" of the form

$$a + p^n\mathbb{Z}_p = D(a, p^{-n}) = \{x \in \mathbb{Q}_p : \|x - a\|_p \leq 1/p^n\}$$

for any center $a \in \mathbb{Q}_p$ and radius p^{-n}.

Lemma 2.4 *If $b \in D(a, p^{-n})$, then $D(a, p^{-n}) = D(b, p^{-n})$.*

In other words, every point of the disc $D(a, p^{-n})$ is a center.

Proof. If $x \in D(a, p^{-n})$, the non-Archimedean property of $\| \ \|_p$ implies that $\|x - b\|_p = \|x - a + a - b\|_p \leq \max\{\|x - a\|_p, \|a - b\|_p\} \leq p^{-n}$, and so $D(a, p^{-n}) \subset D(b, p^{-n})$. The opposite inclusion follows by reversing the roles of a and b in this argument. $\square$

It then follows that if two discs have a nontrivial intersection, one of them must be contained in the other. We thus see that all of the basis elements of the topology of $\mathbb{Q}_p$ are also closed sets; for given any $a \in \mathbb{Q}_p$, we have that $\mathbb{Q}_p \setminus (a + p^n \mathbb{Z}_p) = \cup_{\tilde{a} \in \mathbb{Q}_p \setminus (a + p^n \mathbb{Z}_p)} (\tilde{a} + p^n \mathbb{Z}_p)$ is a union of basis elements.

Given a subset X of $\mathbb{Q}_p$, we shall sometimes let $D_X(a, p^{-n})$ stand for the relative disc $D(a, p^{-n}) \cap X$.

The ring of integers $\mathbb{Z}_p$ inherits the subspace topology, with its basis of open sets given by $\{(a + p^n \mathbb{Z}_p) \cap \mathbb{Z}_p\}_{a \in \mathbb{Q}_p}$. Since the rational integers are dense in $\mathbb{Z}_p$, and every point of a disc is a center, the countable family $\{(a + p^n \mathbb{Z}_p)\}_{a \in \mathbb{Z}, 0 \le a < p^n}^{n \in \mathbb{Z}}$ constitutes a basis for the topology of $\mathbb{Z}_p$ also.

If n is any nonnegative rational integer, the collection

$$\{D_{\mathbb{Z}_p}(a, p^{-n})\}_{a=0}^{p^n - 1}$$

forms a covering of $\mathbb{Z}_p$ by discs that are pairwise disjoints. Below we shall use this fact on several occasions.

We have observed the compactness of $\mathbb{Z}_p$ earlier, when we derived the p-adic integers as a projective limit. Since that argument makes use of Tychonoff's theorem, we pause to provide a simpler proof of the compactness of $\mathbb{Z}_p$ proceeding directly, from the definition.

Proposition 2.5   *The space $\mathbb{Z}_p$ is compact. The space $\mathbb{Q}_p$ is locally compact.*

Proof. We need to prove that any covering of $\mathbb{Z}_p$ has a finite subcovering. Since open sets are unions of discs, it suffices to show that an arbitrary covering of $\mathbb{Z}_p$ by discs $\{D(a_i, p^{-n_i})\}_{a_i, n_i \in \mathbb{Z}, 0 \le a_i < p^{n_i}}^{i \in I}$, where I is some index set, has a finite subcovering. We prove that there exists a finite subset $\tilde{I}$ of I such that $\{D(a_i, p^{-n_i})\}_{i \in \tilde{I}}$ covers $\mathbb{Z}_p$ also.

If at least one of the n_is is less or equal than zero, the radius of the corresponding disc would be greater or equal than one, and by Lemma 2.4, that single disc would suffice to cover $\mathbb{Z}_p$. Thus, we only need to consider the case where $n_i > 0$ for each $i \in I$. Under this assumption, let $\bar{n} = \inf\{n_i : i \in I\}$. Then, there exists at least one disc in the given covering of radius $p^{-\bar{n}}$.

For each rational integer a, let n_a be an integer such that a is contained in a disc of the given covering of radius $1/p^{n_a}$, and define

$$N = \max\{n_a : 0 \le a \le p^{\bar{n}} - 1\}.$$

We know that

$$\mathbb{Z}_p = \cup_{j=0}^{p^N - 1} D_{\mathbb{Z}_p}(j, p^{-N}),$$

and by Lemma 2.4 and the remark immediately after its proof, we conclude that for each of the discs $D(j, p^{-N})$ there exists an index $i_j \in I$ such that $D(j, p^{-N}) \subset D(a_{i_j}, p^{-n_{i_j}})$. If $\tilde{I} = \{i_0, \ldots, i_{p^N - 1}\}$, we then have that

$$\mathbb{Z}_p \subset \cup_{i \in \tilde{I}} D(a_i, p^{-n_i}),$$

proving the desired result.

That $\mathbb{Q}_p$ is locally compact follows since the linear mapping $x \mapsto a + p^n x$ is a homeomorphism between $\mathbb{Z}_p$ and the disc $D(a, p^{-n})$, respectively. $\square$

Remark 2.6 Compactness of $\mathbb{Z}_p$ can be proven in yet another way, by showing that $\mathbb{Z}_p$ is sequentially compact, which on metrizable spaces is equivalent. Indeed, given any sequence $\{\alpha_n\}$ in $\mathbb{Z}_p$, we can construct a subsequence all of whose elements have the same first digit in their p-adic expansions (2.4), and by iteration of this argument k times, find a subsequence $\{\alpha_{n_j}^k\}_{j=1}^{\infty}$ such that the first k digits in the p-adic expansions of all the $\alpha_{n_j}^k$ s coincide. We then "diagonalize" to define a subsequence $\{\alpha_{n_j}^j\}$ of $\{\alpha_n\}$ that converges by choosing the first element of the first subsequence, the second of the second, and so on. $\square$

Lemma 2.7 *The set of p-adic numbers $\mathbb{Q}_p$ is totally disconnected.*

Proof. Let S be any subset with more than one point. If x and y are two elements of S, then we have that $\|x - y\|_p > 0$, and if $D = \{z \in S : \|z - x\|_p < \|x - y\|_p\}$, then the pair of open sets D, $S \setminus D$ forms a separation of S. $\square$

2.4 Analytic and algebraic properties of $\mathbb{Q}_p$

Let p and $\tilde{p}$ be two distinct primes. Since the sequence $\{p^n\}$ converges in $\mathbb{Q}_p$ but not in $\mathbb{Q}_{\tilde{p}}$, as topological fields, $\mathbb{Q}_p$ and $\mathbb{Q}_{\tilde{p}}$ are not isomorphic. We show here that, in fact, they are not even isomorphic as fields. This shall follow by a simple analysis of solutions to quadratic equations.

Let us consider the polynomial $f(x) = x^2 - a$, where $a \in \mathbb{Z}$ is not divisible by p^2. If p divides a then f cannot have a root in $\mathbb{Q}_p$. For if $\alpha = \sqrt{a}$ is in $\mathbb{Q}_p$, since $\|a\|_p = \|\sqrt{a}\|_p \|\sqrt{a}\|_p$, we must have that $\|\sqrt{a}\|_p = \|p\|_p^{\frac{1}{2}}$, which is not possible because $\| \ \|_p$ ranges over integer powers of $\|p\|_p$.

If p does not divide a and we can find a p-adic number α such that

$\|f(\alpha)\|_p < 1$, by the congruences (2.6) for α, we can find $a_1 \in \mathbb{Z}$ such that

$$a_1^2 \equiv a \bmod p, \tag{2.8}$$

which says that a is a quadratic residue mod p. If p is an odd prime, this equation has two solutions in the range $1 \le a_1 \le p - 1$. Proceeding by induction, we find two sequences of rational integers $\{a_n\}$ that converge to p-adic numbers $\sqrt{a}$ such that $\sqrt{a} \equiv a_n \bmod p^n$, and where the starting element a_1 is equal to each one of the said solutions of the congruence (2.8), respectively. Thus, f has two distinct roots in this case.

If a is a quadratic nonresidue mod p, then f cannot have a root $\alpha \in \mathbb{Q}_p$. For otherwise, there would exist $c \in \mathbb{Z}$ such that

$$\alpha + p\mathbb{Z}_p = c + p\mathbb{Z}_p,$$

and so,

$$a + p\mathbb{Z}_p = \alpha^2 + p\mathbb{Z}_p = c^2 + p\mathbb{Z}_p,$$

which implies that

$$c^2 \equiv a \bmod p.$$

Thus, a would be a quadratic residue, contrary to the original assumption.

The argument given above is truly built on the p-adic version of Newton's iteration algorithm. We provide some of the details of this in order to explain the extra care that must be exercised in our argument in the case when the prime p is 2.

Indeed, if f is any monic polynomial in $\mathbb{Z}_p[x]$, and we have $a_1 \in \mathbb{Z}$ such that $\|f(a_1)\|_p < 1$, then the sequence given by

$$a_{n+1} = a_n - (f'(a_n))^{-1} f(a_n) \tag{2.9}$$

is expected to converge to a root of f under suitable hypothesis on the p-adic norm of $f'(a_1)$. Indeed, if we assume that $\left\| f'(a_1) \right\|_p = 1$, by the Taylor expansion

$$f(x + h) = f(x) + hf'(x) + h^2 e_f(x, h),$$

we see that

$$f(a_2) = f(a_1) - \frac{f(a_1)}{f'(a_1)} f'(a_1) + \left(\frac{f(a_1)}{f'(a_1)} \right)^2 e_f\left(a_1, -\frac{f(a_1)}{f'(a_1)} \right),$$

where the error $e_f(a_1, -f(a_1)/f'(a_1))$ has p-adic norm bounded above by 1. Then, $\|f(a_2)\|_p \le \|f(a_1)\|_p^2$, and $\|a_2 - a_1\|_p \le \|f(a_1)\|_p$. Proceeding by induction, we see that

$$\|a_{n+1} - a_n\|_p \le \|f(a_1)\|_p^{2^{n-1}}, \quad \|f(a_{n+1})\|_p \le \|f(a_1)\|_p^{2^n}.$$

Thus,

$$a = a_1 + (a_2 - a_1) + (a_3 - a_2) + \cdots = \lim_{n \to \infty} a_n$$

converges p-adically, and the limit is a p-adic integer root of $f(x)$.

This argument must be modified slightly when dealing with the prime $p = 2$ and the quadratic polynomial $f(x) = x^2 - a$. For in this case, $f'(a_1) = 2a_1$ will have relatively small norm, which affects the convergence rate of the iteration scheme. But it does not prevent it. For if $f(a_1)$ has norm less than 1, and $0 \neq \left\| f'(a_1) \right\|_p < 1$, as long as we assume that $f(a_1)/f'(a_1)^2$ has p-adic norm less than 1, the iteration scheme above produces a sequence that converges p-adically to a root of f.

By applying this iteration scheme to our quadratic polynomial for all possible choices of the prime p, we obtain the following result.

Theorem 2.8 *Let a be any rational integer not divisible by p^2. Consider the equation $x^2 - a = 0$. Then:*

a). *If $p \mid a$, the equation has no solution in $\mathbb{Q}_p$.*

b). *If $p \neq 2$ and $p \nmid a$, the equation has two solution in $\mathbb{Q}_p$ if a is a quadratic residue mod p, and no solution if it is not.*

c). *If $p = 2$ and $2 \nmid a$, the equation has two solutions in $\mathbb{Q}_p$ if a is congruent to 1 mod 8, and no solution if it is not.*

We now reconsider the distinct primes p and $\tilde{p}$ of the beginning, with say, $\tilde{p} < p$. If these primes are both odd, and p is a quadratic residue mod $\tilde{p}$, the equation $x^2 - p = 0$ has two solutions in $\mathbb{Q}_{\tilde{p}}$, but none in $\mathbb{Q}_p$. If, on the other hand, p is a quadratic nonresidue mod $\tilde{p}$, we may choose a quadratic nonresidue n mod $\tilde{p}$ with $n < \tilde{p}$. Then np is a quadratic residue mod $\tilde{p}$, and the equation $x^2 - np = 0$ has two solutions in $\mathbb{Q}_{\tilde{p}}$ but none in $\mathbb{Q}_p$.

The case where p is an odd prime and $\tilde{p} = 2$ is treated similarly. If 2 is a quadratic residue mod p, the equation $x^2 - 2 = 0$ has two solutions in $\mathbb{Q}_p$ but none in $\mathbb{Q}_2$. If 2 is a quadratic nonresidue mod p, let n be an odd quadratic nonresidue mod p. Then the equation $x^2 - 2n = 0$ has solutions in $\mathbb{Q}_p$ but none in $\mathbb{Q}_2$.

We thus conclude the following.

Theorem 2.9   *If p and $\tilde{p}$ are distinct primes, the fields $\mathbb{Q}_p$ and $\mathbb{Q}_{\tilde{p}}$ are not isomorphic.*

Remark 2.10   The Newton's iteration algorithm on $\mathbb{Z}_p$ discussed above is commonly known as *Hansel's lemma*. A convenient formulation that can be used to find roots of polynomial equations in several variables goes as follows: *Let $F(x) = (f_1(x), \ldots, f_n(x))$ with $f_j(x) = f_j(x_1, \ldots, x_n) \in \mathbb{Z}_p[x_1, \ldots, x_n]$, $j = 1, \ldots, n$. Let $a_1 \in \mathbb{Z}_p^n$ be such that $F(a_1) \equiv 0 \mod p$ and such that the Jacobian matrix $F'(a_1) = F'(x)\,|_{x=a_1}$ is invertible. If $\|F'(a_1)\|_p \leq 1$ and $\|F'(a_1)\|_p^{-1} \left\|(F'(a_1))^{-1}F(a_1)\right\|_p < 1$, then there exists a unique $a \in \mathbb{Z}_p^n$ such that $F(a) = 0$ and $a \equiv a_1 \mod p$.*

The proof follows along the lines outlined above in the case of a single variable, where starting from the approximate solution a_1, the elements in the sequence (2.9) brings us closer to the actual root at every step.

As an example, let us consider the quadratic polynomial $f(x, y, z) = 3x^2 + y^2 + 4z^2 + 2(xy + xz + yz)$, and let p be the prime 7. For $a_1 = (0, -1, -1) \in \mathbb{Z}_7^3$ we have that $f(a_1) = 7 \equiv 0 \mod 7$ and $f'(a_1) = (-4, -4, -10)$. In order to analyze the roots of f by way of the stated result, we consider the polynomial function $F : \mathbb{Z}_7^3 \to \mathbb{Z}_7^3$ given by $F(x, y, z) = (f(x, y, z), y - 6, z - 6)$. Then the sequence (2.9) yields $a_2 = a_1 + (F'(a_1))^{-1}F(a_1) = (-91/4, 6, 6)$, and we have that $F(a_2) = (3 \cdot 7^2 \cdot 137/2^4, 0, 0)$. Now we set $a_3 = a_2 - (F'(a_2))^{-1}F(a_2) = (-6, 937/600, 6, 6)$, and we have that $F(a_3) = (7^4 \cdot 137^2/(2^6 \cdot 3 \cdot 5^4), 0, 0)$. Iterating this procedure, we obtain a Cauchy sequence $\{a_j\}$ in $\mathbb{Z}_7^3$ such that $\|F(a_j)\|_7 \leq 1/7^j$, and the continuity of polynomial functions will imply that the limit a in $\mathbb{Z}_7^3$ is a root of F, and so a root of its first coordinate function f. By construction, $a \equiv a_1 \mod 7$. $\square$

2.5 $(p-1)$-roots of unity in $\mathbb{Q}_p$

The field $\mathbb{Q}_p$ contains all $(p-1)$-roots of unity. Indeed, the equation

$$x^{p-1} - 1 = 0$$

has exactly $p-1$ roots in $\mathbb{Q}_p$, the maximum number of possible solutions, as we now see.

By Euler's theorem, given any integer a in the range from 1 to $p-1$, we have that $a^{\varphi(p^{n+1})} = a^{p^n(p-1)} \equiv 1 \mod p^{n+1}$. Thus, the series

$$\alpha_a = \lim a^{p^n} = a + (a^p - a) + (a^{p^2} - a^p) + \cdots$$

converges. For using the non-Archimedean property of $\| \ \|_p$, it suffices to show that

$$a^{p^{n+1}} - a^{p^n} = a^{p^n}\left(a^{p^n(p-1)} - 1\right),$$

goes to zero as n goes to ∞, which is so as the last factor on the right of this expression is divisible by p^{n+1}. We also have that

$$\alpha_a^{p-1} = \lim a^{p^n(p-1)} = 1,$$

and so α_a is a $(p-1)$-root of unity in $\mathbb{Q}_p$.

If two of these roots were to coincide,

$$\alpha_a + p\mathbb{Z}_p = \alpha_{\tilde{a}} + p\mathbb{Z}_p,$$

we would have that $a \equiv \tilde{a} \bmod p$, and so a would have to be equal to $\tilde{a}$. Thus, the collection of p-adic numbers $\{\alpha_a\}_{a=1}^{p-1}$ exhausts the set of all $(p-1)$-roots of unity.

We summarize our discussion in the form of a theorem. We let $\mathbb{Z}_p^\times$ denote the subgroup of p-adic units in $\mathbb{Z}_p$, and let $\mathbb{F}_p$ be a canonical realization of a finite field with p-elements. We denote by $\mathbb{F}_p^\times$ the multiplicative group of nonzero elements of $\mathbb{F}_p$.

Theorem 2.11 *Consider $1 + p\mathbb{Z}_p$ as a multiplicative subgroup of $\mathbb{Z}_p$, and let $\mu(\mathbb{Z}_p)$ denote the multiplicative subgroup of roots of unity, $\{x \in \mathbb{Z}_p : x^{p-1} = 1\}$. Then we have that $\mathbb{Z}_p^\times = \mu(\mathbb{Z}_p) \times (1 + p\mathbb{Z}_p)$, and $\mu(\mathbb{Z}_p)$ is the unique subgroup of $\mathbb{Z}_p^\times$ isomorphic to $\mathbb{F}_p^\times$.*

We have the field identifications $\mathbb{Z}_p/p\mathbb{Z}_p \cong \mathbb{Z}/p\mathbb{Z} \cong \mathbb{F}_p$.

Let us now recall that for a finite field $\mathbb{F}_q$ with q elements and characteristic p, the Frobenius mapping is given by

$$\mathbb{F}_q \xrightarrow{\phi} \mathbb{F}_q$$
$$x \mapsto x^p.$$

It defines the unique automorphism of $\mathbb{F}_q$ that fixes the subfield $\mathbb{F}_p$, and is a generator of $\mathrm{Aut}(\mathbb{F}_q)$. Since $x^p = x$ for any $x \in \mathbb{F}_p$, the Frobenius mapping reduces to the identity when $q = p$. This fact implies that the identity is the unique lift to $\mathbb{Z}_p$ of the p-th power Frobenius isomorphism on $\mathbb{F}_p \subset \mathbb{Z}_p$, and in turn, we may now derive the following simple consequence, which will be of crucial importance to us later on, when we in fact introduce the arithmetic differential operators on $\mathbb{Z}_p$. The reader may find our discussion a bit odd at this point, since after all, the identity is the unique ring automorphism on $\mathbb{Z}_p$. This oddity will be clarified later, and the reader should merely keep in mind now that the study arithmetic differential operators can be done with more generality [8, 6, 15] over rings larger than $\mathbb{Z}_p$, and where the lift of the p-th power Frobenius isomorphism is not the identity.

Theorem 2.12 *For all $x \in \mathbb{Z}_p$, we have that $x - x^p \in p\mathbb{Z}_p$.*

Proof. We have that $x = \alpha_x + pu$ for elements u and α_x in $\mathbb{Z}_p$, α_x a $p - 1$ root of unity. Since $\begin{pmatrix} p \\ j \end{pmatrix} \equiv 0 \bmod p$ for $1 \leq j < p$, we then have that $x^p = \alpha_x^p + p\mathbb{Z}_p = \alpha_x + p\mathbb{Z}_p = x + p\mathbb{Z}_p$, and the desired result follows. $\square$

3

Some classical analysis on $\mathbb{Q}_p$

The convergence of a p-adic series $\sum_{n=0}^{\infty} a_n$ is somewhat easy to analyze. The convergence of the series implies that $\lim_{n \to \infty} a_n = 0$, but the remarkable fact is that this condition alone implies the converse. For if $s_n = \sum_{j=0}^{n} a_j$ is the sequence of partial sums, by the non-Archimedean property of the norm we have that

$$\|s_n - s_m\|_p = \|a_n + \cdots + a_{m+1}\|_p \leq \max_{m+1 \leq j \leq n} \|a_j\|_p \to 0\,,$$

showing that $\{s_n\}$ is a Cauchy sequence, and so its limit exists in $\mathbb{Q}_p$. Using this same argument, it follows that if $r = 1/\limsup \|a_j\|_p^{1/j}$, the power series

$$F(x) = \sum_{n=0}^{\infty} a_n x^n \in \mathbb{Q}_p[[x]] \tag{3.1}$$

converges if $\|x\|_p < r$, diverges if $\|x\|_p > r$, and the case $\|x\|_p = r$ is decided by simply looking at $\lim \|a_j\|_p r^j$, which implies convergence or divergence contingent upon it being zero or not.

Let us assume that the series $F(x)$ in (3.1) converges, and so we have that $\lim_{n \to \infty} a_n x^n = 0$. Let $F'(x)$ denote the formal power series of term-by-term derivatives,

$$F'(x) = \sum_{n=0}^{\infty} n a_n x^{n-1}\,.$$

Since $\|n a_n x^{n-1}\|_p \leq \|a_n x^{n-1}\|_p \to 0$, the series $F'(x)$ converges also. Furthermore, $F'(x)$ is the limit of the usual quotient of increments

23

$(F(x+y) - F(x))/y$. For, by the binomial expansion, we have that

$$\frac{F(x+y) - F(x)}{y} = \sum_{n=1}^{\infty} a_n \sum_{k=1}^{n} \binom{n}{k} x^{n-k} y^{k-1},$$

and if $\|y\|_p < \|x\|_p$, we have

$$\left\| a_n \binom{n}{k} x^{n-k} y^{k-1} \right\|_p \leq \|a_n\|_p \|x\|_p^{n-1} \to 0.$$

Hence, the series above converges uniformly in y so long as y is p-adically smaller than x, and the infinite summation and limit as $y \to 0$ can be intertwined. We obtain

$$
\begin{aligned}
F'(x) = \lim_{y \to 0} \frac{F(x+y) - F(x)}{y} \ &= \ \sum_{n=1}^{\infty} a_n \lim_{y \to 0} \sum_{k=1}^{n} \binom{n}{k} x^{n-k} y^{k-1} \\
&= \ \sum_{n=1}^{\infty} n a_n x^{n-1} = F'(x),
\end{aligned}
$$

as desired.

By iteration we conclude that a convergent power series has convergent power series derivatives of all order.

In a sense to be made precise below, the polynomial functions $x \mapsto \binom{x}{n}$ play the role in p-adic analysis that the monomials x^n do in real analysis. We lead to the discussion of this assertion, a theorem of Mahler [36], by first showing the following result (see [2]).

Theorem 3.1   *Let* $y \in \mathbb{Q}_p$ *be a* p-*adic number such that* $\mathrm{ord}_p\, y \geq 0$. *Then the binomial series*

$$(1+x)^y = \sum_{n=0}^{\infty} \binom{y}{n} x^n$$

converges for all x *with* $\mathrm{ord}_p\, x > 1/(p-1)$.

Notice that, for all p other than 2, $1/(p-1) \notin \mathbb{Z}$, and so the lower bound we give above for the p-adic order of x appears to be of a rather peculiar nature. The reason for doing so is simple. The binomial series function can be defined also on the completion of the algebraic closure of $\mathbb{Q}_p$, a rather large field to which the p-adic norm and order admit extensions (see §3.2 below). The function there defined will converge precisely under the condition stated in the theorem.

Proof. We have

$$\left(\begin{array}{c} y \\ n \end{array} \right) = \frac{y(y-1)\cdots(y-n+1)}{n!} .$$

Since the p-adic order of y is nonnegative, the p-adic order of $\left(\begin{array}{c} y \\ n \end{array} \right)$ is bounded below by that of $1/n!$, and we conclude that

$$\mathrm{ord}_p \left(\begin{array}{c} y \\ n \end{array} \right) x^n \geq \mathrm{ord}_p \frac{x^n}{n!} .$$

Let $n = a_0 + a_1 p + \cdots + a_k p^k$ be the base p expansion of n. We define $s_n = a_0 + \cdots + a_k$, and set $s_0 = 0$. The p-adic order of n is the smallest integer l such that $a_l \neq 0$, and we see easily that $s_{n-1} = l(p-1) + s_n - 1$, so $\mathrm{ord}_p n = (s_{n-1} - s_n + 1)/(p-1)$. By the multiplicative property of ord_p, it follows that

$$\mathrm{ord}_p n! = \frac{n - s_n}{p - 1} . \tag{3.2}$$

Hence,

$$\begin{aligned} \mathrm{ord}_p \frac{x^n}{n!} &= n \, \mathrm{ord}_p x - \mathrm{ord}_p n! \\ &= n \, \mathrm{ord}_p x - \frac{n - s_n}{p - 1} \\ &= n \left(\mathrm{ord}_p x - \frac{1}{p-1} \right) + \frac{s_n}{p - 1} , \end{aligned} \tag{3.3}$$

which grows with no bounds under the stated hypothesis, and so the n-th term of the series defining $(1 + x)^y$ converges p-adically to zero. The result follows. $\qquad\square$

We may show that the function of y that we obtained above is continuous. For all y in $\mathbb{Z}_p$, the argument given shows that if x has p-adic order greater than $1/(p-1)$, we have $\left\| \left(\begin{array}{c} y \\ n \end{array} \right) x^n \right\|_p \leq \| x^n / n! \|_p \to 0$, and the series defining $(1 + x)^y$ converges uniformly as a function of y. But uniform convergence of continuous functions yield a continuous limit. So the function $(1 + x)^y$ is continuous as function of $y \in \mathbb{Z}_p$. We address this type of issues in further detail now.

We restrict our attention to the space $C(\mathbb{Z}_p, \mathbb{Q}_p)$ of $\mathbb{Q}_p$-valued continuous function on $\mathbb{Z}_p$. Since $\mathbb{Z}_p$ is compact, any $f \in C(\mathbb{Z}_p, \mathbb{Q}_p)$ is uniformly continuous. If we were to consider $C(D, \mathbb{Q}_p)$ instead, where $D = D(a, p^{-n})$ is the disc in $\mathbb{Q}_p$ of radius p^{-n} centered at a, a similar

result to the one below would still hold since $D(a, p^{-n})$ is homeomorphic
to $\mathbb{Z}_p$.

Since the nonnegative rational integers form a dense subset of $\mathbb{Z}_p$, a
function $f \in C(\mathbb{Z}_p, \mathbb{Q}_p)$ is completely determined by its values on the
set $\{0, 1, 2, \ldots\}$.

Given any $f \in C(\mathbb{Z}_p, \mathbb{Q}_p)$, let us consider the sequence

$$a_n = \sum_{k=0}^{n} (-1)^k \binom{n}{k} f(n-k) \in \mathbb{Q}_p , \qquad (3.4)$$

and the associated power series

$$F(x) = \sum_{n=0}^{\infty} a_n \binom{x}{n} \in \mathbb{Q}_p[[x]] . \qquad (3.5)$$

For any nonnegative rational integer m, $\binom{m}{n} = 0$ if $n > m$. Then

$$F(m) = \sum_{n=0}^{m} a_n \binom{m}{n} = \sum_{l=0}^{m} f(l) \sum_{k=0}^{m-l} (-1)^k \binom{k+l}{l} \binom{m}{k+l} .$$

Since

$$\binom{k+l}{l} \binom{m}{k+l} = \binom{m}{l} \binom{m-l}{k} ,$$

we obtain

$$
\begin{aligned}
F(m) &= \sum_{l=0}^{m} \binom{m}{l} f(l) \sum_{k=0}^{m-l} (-1)^k \binom{m-l}{k} \\
&= \sum_{l=0}^{m} \binom{m}{l} f(l)(1-1)^{m-l} \\
&= f(m) .
\end{aligned}
$$

Thus, the power series (3.5) interpolates the values of the function f over
the dense subset of nonnegative rational integers. Therefore, if we were to
prove that this series converges and defines an element of $C(\mathbb{Z}_p, \mathbb{Q}_p)$, the
said element will have to be $f(x)$ itself. We proceed to prove exactly that
in a slightly oblique but succinct manner. The end result is a theorem
due to K. Mahler [36].

Let us notice that the polynomial functions $x \mapsto \binom{x}{n}$ map rational
integers to rational integers, and so define elements of $C(\mathbb{Z}_p, \mathbb{Z}_p)$. In
what follows, we shall say that a sequence $\{a_n\}_{n=0}^{\infty} \subset \mathbb{Q}_p$ is restricted if
$a_n \to 0$ p-adically.

Theorem 3.2   (Mahler [36].) *Let* $\mathcal{S} = \{\{a_n\}_{n=0}^{\infty} \subset \mathbb{Q}_p \,, \; a_n \to 0\}$ *be the set of restricted sequences. The mapping*

$$
\begin{aligned}
\mathcal{S} &\longrightarrow C(\mathbb{Z}_p, \mathbb{Q}_p) \\
\{a_n\} &\mapsto \sum_{n=0}^{\infty} a_n \begin{pmatrix} x \\ n \end{pmatrix}
\end{aligned}
\tag{3.6}
$$

is a bijection.

Given a restricted series $\{a_n\}$, the mapping (3.6) in this statement is given by the associated series (3.5). We show first that this mapping is well defined.

The elementary combinatorial identity

$$
\begin{pmatrix} x \\ n \end{pmatrix} = \sum_{k=0}^{n} \begin{pmatrix} x-y \\ k \end{pmatrix} \begin{pmatrix} y \\ n-k \end{pmatrix}
$$

holds for p-adic integers x, y also. This can be shown by using Theorem 3.1 to determine the coefficient of t^n in the series expansion of the identity $(1+t)^x = (1+t)^{x-y}(1+t)^y$. Given an arbitrary integer x in $\mathbb{Z}_p$, let y be a nonnegative rational integer such that

$$
\left\| \frac{x-y}{n!} \right\|_p \leq 1 .
$$

We have

$$
\left\| \begin{pmatrix} y \\ n-k \end{pmatrix} \right\|_p \leq 1 ,
$$

as the argument to the function $\| \; \|_p$ on the left side is just the standard binomial coefficient. On the other hand,

$$
\begin{pmatrix} x-y \\ k \end{pmatrix} = \frac{(x-y)(x-y-1)\cdots(x-y-k+1)}{k!} = \frac{x-y}{k!} u_k \,,
$$

where $u_k \in \mathbb{Z}_p$, and therefore,

$$
\left\| \begin{pmatrix} x-y \\ k \end{pmatrix} \right\|_p \leq 1 .
$$

By the non-Archimedean property of the p-adic norm, we conclude then that

$$
\left\| \begin{pmatrix} x \\ n \end{pmatrix} \right\|_p = \left\| \sum_{k=0}^{n} \begin{pmatrix} x-y \\ k \end{pmatrix} \begin{pmatrix} y \\ n-k \end{pmatrix} \right\|_p \leq 1 ,
$$

estimate that implies that the series in the statement of the theorem defines a continuous function, and so the map (3.6) is well-defined. Indeed, the series converges because the said estimate implies that

$$\left\| a_n \begin{pmatrix} x \\ n \end{pmatrix} \right\|_p \le \|a_n\|_p \to 0 \,.$$

On the other hand, the sequence of partial sums are polynomials, hence continuous, and the limit, being uniform, must be also a continuous function.

The injectivity of (3.6) can now be argued with ease. Indeed, let us consider a restricted sequence $\{a_n\}$ such that its associated power series $F(x)$ in (3.5) is the function identically zero. By the combinatorial identity

$$\begin{pmatrix} x+1 \\ n \end{pmatrix} = \begin{pmatrix} x \\ n \end{pmatrix} + \begin{pmatrix} x \\ n-1 \end{pmatrix}$$

that holds for any p-adic integer x also, we have that

$$0 = F(x+1) - F(x) = \sum_{n=1}^{\infty} a_n \begin{pmatrix} x \\ n-1 \end{pmatrix} \,.$$

We now observe that the value of the continuous function $F(x+1) - F(x)$ at $x = n \in \mathbb{Z}_{\ge 0}$ is equal to a_{n+1}, and so proceeding by induction, we conclude that $a_n = 0$ for all $n \ge 1$. So $F(x)$ reduces to the zero term a_0, and this must be zero also since, by assumption, so is F.

In order to finish the proof of Theorem 3.2, we are left with the task of showing the surjectivity of (3.6). So let us take $f \in C(\mathbb{Z}_p, \mathbb{Q}_p)$. By compactness of $\mathbb{Z}_p$, f is bounded and so, after multiplication by a suitable power of p, we obtain a function that ranges in $\mathbb{Z}_p$. Thus, without loss of generality, we may assume that f is an element of $C(\mathbb{Z}_p, \mathbb{Z}_p)$. We view the latter as a Banach space with the supremum norm.

By completeness of $C(\mathbb{Z}_p, \mathbb{Z}_p)$, the desired surjectivity follows if we show that for any $N \ge 0$ there exist $\{a_n\} \in \mathbb{Z}_p$ and $f_N \in C(\mathbb{Z}_p, \mathbb{Z}_p)$ such that

$$f(x) = \sum_n a_n \begin{pmatrix} x \\ n \end{pmatrix} + p^N f_N(x) \,.$$

We argue this by induction on N, starting with $f_0 = f$, cf. [34], pp. 99-100.

Notice that it suffices to prove that f_N is, modulo p, a linear combination of the $\begin{pmatrix} x \\ n \end{pmatrix}$s. By the uniform continuity of f_N, there exists M

such that

$$\|f_N(x) - f_N(y)\|_p \leq \frac{1}{p} \quad \text{for all } x, y \in \mathbb{Z}_p \text{ such that } \|x - y\|_p \leq \frac{1}{p^M}.$$

Hence, the map

$$\mathbb{Z}_p \xrightarrow{f_N} \mathbb{Z}_p \to \mathbb{F}_p = \mathbb{Z}_p/p\mathbb{Z}_p$$

factors through

$$\mathbb{Z}/p^M\mathbb{Z} \to \mathbb{F}_p.$$

It is rather elementary to see that for any k in the range $0 \leq k < p^M$, the map

$$\mathbb{Z}_p \xrightarrow{\binom{x}{k}} \mathbb{Z}_p \to \mathbb{F}_p$$

factors through

$$\mathbb{Z}/p^M\mathbb{Z} \to \mathbb{F}_p$$

also. Since $\mathbb{F}_p$ is discrete and finite, the desired task will be accomplished if we show that the set of tuples $\{(\bar{a}_0, \ldots, \bar{a}_{p^M-1}) : \bar{a}_i \in \mathbb{F}_p\}$, and the set of mappings $\text{Maps}(\mathbb{Z}/p^M\mathbb{Z}, \mathbb{F}_p)$ are in 1-to-1 correspondence with each other via the function

$$(\bar{a}_0, \ldots, \bar{a}_{p^M-1}) \mapsto \sum_{n=0}^{p^M-1} \bar{a}_n \binom{x}{n}.$$

The said sets have the same number of elements. On the other hand, we can prove the injectivity of the map above by an argument similar to that used to prove the injectivity of (3.6) itself. This completes the proof of the Theorem. $\qquad\square$

Example 3.3 Consider the continuous function $\mathbb{Z}_5 \ni y \mapsto (1 - 5)^y$ given by Theorem 3.1. We have that

$$(1 - 5)^y = \sum_{n=0}^{\infty} (-5)^n \binom{y}{n},$$

where the series on the right converges in $\mathbb{Z}_5$. $\qquad\square$

3.1 The Artin–Hasse exponential function

Using (3.2) we derived identity (3.3). It follows that if $\operatorname{ord}_p x > 1/(p-1)$, $\operatorname{ord}_p (x^n/n!) \to \infty$, and the exponential series

$$e^x = \sum_{n=0}^{\infty} \frac{x^n}{n!}$$

converges for these xs. If otherwise x is such that $\operatorname{ord}_p x \le 1/(p-1)$, we may take $n = p^k$. We then have that $s_n = 1$, and $\operatorname{ord}_p (x^n/n!) \not\to \infty$, so the series above does not converge p-adically in this case. Thus, the exponential function has p-adic radius of convergence equal to $p^{-\frac{1}{p-1}}$. Bigger and bigger denominators in the coefficients of the series help its convergence when we use the classical Euclidean norm, but the opposite is true p-adically when the denominators in question are divisible by larger and larger powers of p. This makes the p-adic radius of convergence of the exponential function relatively small compared to its radius of convergence over $\mathbb{Q}_\infty$.

Let $\mu(n)$ be the Möbius function, defined by

$$\mu(n) = \begin{cases} (-1)^k & \text{if } n \text{ is the product of } k \text{ distinct primes,} \\ 0 & \text{otherwise}. \end{cases}$$

We recall that by the prime factorization theorem, we have that

$$\sum_{d|n} \mu(d) = 0 \text{ for } n > 1. \tag{3.7}$$

Lemma 3.4 *We have the identity*

$$e^x = \prod_{n=1}^{\infty} (1 - x^n)^{-\frac{\mu(n)}{n}}.$$

Proof. Taking log on the right side, we obtain that

$$-\sum_{n=1}^{\infty} \frac{\mu(n)}{n} \log(1 - x^n) = \sum_{n=1}^{\infty} \frac{\mu(n)}{n} \sum_{m=1}^{\infty} \frac{x^{nm}}{m} = \sum_{j=1}^{\infty} \frac{x^j}{j} \sum_{d|j} \mu(d) = x,$$

where the last equality follows by (3.7). $\qquad\square$

The identity in Lemma 3.4 shows that the failure for the series defining e^x to converge p-adically arises from the terms where n is divisible by p and square-free. This leads to the following natural definition:

Definition 3.5 Given a prime p, the Artin–Hasse exponential function is defined by

$$E_p(x) = \prod_{\substack{n=1 \\ p \nmid n}}^{\infty} (1 - x^n)^{-\frac{\mu(n)}{n}}.$$

$\square$

It follows that $E_p(x) \in 1 + x\mathbb{Z}_p[[x]]$, and therefore, $E_p(x)$ converges for $\|x\|_p < 1$.

As in the proof of Lemma 3.4, we may apply the log function to the infinite product defining $E_p(x)$. If we then use the identity

$$\sum_{\substack{d|n \\ p \nmid d}} \mu(d) = \begin{cases} 1 & \text{if } n \text{ is a power of } p, \\ 0 & \text{otherwise}, \end{cases} \tag{3.8}$$

we see that

$$E_p(x) = \exp\left(\sum_{j=0}^{\infty} \frac{x^{p^j}}{p^j}\right). \tag{3.9}$$

The expression above may be used as an alternative definition for $E_p(x)$, in which case this function will be the exponential of a power series in $\mathbb{Q}[[x]]$. In so doing, it becomes harder to conclude that $E_p(x)$ is in effect an element of $\mathbb{Z}_p[[x]]$, as the coefficients of the series whose exponential yields $E_p(x)$ have norms that grow without bound in $\mathbb{Q}_p$. We address this issue next, and derive the p-adic integrality property of the coefficients of the said series as a consequence of a beautiful and remarkable result in p-adic analysis due to Dwork, which we take the opportunity to state and prove. In spirit, this result is fundamental for the theory of arithmetic differential equations. It studies the interplay between the p-th power of the function of a number and the function of the p-th power of the number.

We begin this line of reasoning by first showing the relationship between $E_p(x^p)$ and $E_p(x)^p$.

Lemma 3.6 *We have the identity*

$$E_p(x^p) = E_p(x)^p \exp(-px).$$

Proof. Let us set $L(x) = \sum_{j=0}^{\infty} \frac{x^{p^j}}{p^j}$. Then we have that

$$E_p(x) = \exp(L(x)),$$

and the desired identity amounts to showing that $L(x^p) = pL(x) - px$, which is clear. $\qquad\square$

By Lemma 3.6, we see that

$$\frac{E_p(x^p)}{E_p(x)^p} = \exp\left(-px\right) = \sum_{n=0}^{\infty} (-1)^n \frac{p^n}{n!} x^n = 1 + px \sum_{n=1}^{\infty} (-1)^n \frac{p^{n-1}}{n!} x^n.$$

Now by (3.2), we see that

$$\mathrm{ord}_p\left(\frac{p^{n-1}}{n!}\right) = n - 1 - \frac{n - s_n}{p - 1} = \frac{(n-1)(p-2) + s_n - 1}{p - 1},$$

which is nonnegative. Hence, $\exp\left(-px\right) \in 1 + px\mathbb{Z}_p[[x]]$, and we have that

$$\frac{E_p(x^p)}{E_p(x)^p} \in 1 + px\mathbb{Z}_p[[x]]. \tag{3.10}$$

In other words, modulo $p\mathbb{Z}_p[[x]]$, the rational series $E_p(x)$ in (3.9) commutes with the p-th power map. This is amenable for the application of the following result.

Lemma 3.7 (Dwork's lemma) *Let* $f(x) = \sum a_i x^i \in 1 + x\mathbb{Q}_p[[x]]$. *Then* $f(x) \in 1 + x\mathbb{Z}_p[[x]]$ *if, and only if,* $f(x^p)/(f(x))^p \in 1 + px\mathbb{Z}_p[[x]]$.

This can be stated in a more general context (cf. Lemma 2.4 in [15]), but that will not be used in here.

Proof. If $f(x) \in 1 + x\mathbb{Z}_p[[x]]$ and since $(a + b)^p \equiv a^p + b^p \bmod p$, by Theorem 2.12 we have that $f(x)^p = f(x^p) + pg(x)$ where $g(x) \in x\mathbb{Z}_p[[x]]$. Then $f(x^p)/(f(x))^p = 1 - p(g(x)/f(x)^p)$, which belongs to $1 + px\mathbb{Z}_p[[x]]$ because $f(x)^p \in 1 + px\mathbb{Z}_p[[x]]$ has an inverse in the multiplicative group $1 + px\mathbb{Z}_p[[x]]$.

Conversely, let us assume that $f(x^p) = (f(x))^p g(x)$ with $g(x) = \sum b_i x^i \in 1 + px\mathbb{Z}_p[[x]]$. Since by hypothesis the leading coefficient is 1 for both, f and g, we have that

$$1 + \sum_{i=1}^{\infty} a_i x^{pi} = \left(1 + \sum_{i=1}^{\infty} a_i x^i\right)^p \left(1 + \sum_{i=1}^{\infty} b_i x^i\right). \tag{3.11}$$

with the b_is are all rational integers congruent to zero mod p. We use induction to show that $a_i \in \mathbb{Z}_p$ for all i.

Let us assume that a_i is a rational integer for all $i < n$. If $p \nmid n$, the coefficient of x^n in the right side of (3.11) is given by an expression of the form

$$pa_n + R(a_1, \ldots, a_{n-1}, b_1, b_2, \ldots, b_n)$$

where $R(a, b)$ is a p-adic integer that depends nonlinearly on a and linearly on b. Since b is congruent to zero mod p, we have that $R(a, b) \in p\mathbb{Z}_p$. By comparison with the left side of (3.11), this coefficient must be equal to zero, and so we have

$$a_n = -\frac{1}{p}R(a_1, \dots, a_{n-1}, b_1, b_2, \dots, b_n) \in \mathbb{Z}_p \,,$$

as desired. If on the other hand, $p \mid n$, the coefficient of x^n in the right side of (3.11) is given by an expression of the form

$$pa_n + a^p_{n|p} + R(a_1, \dots, a_{n-1}, b_1, b_2, \dots, b_n) \,,$$

where $R(a, b)$ is a p-adic integer in $p\mathbb{Z}_p$ for the same reasons as before. In this case, by comparison with the left side, we have that

$$a_{n|p} = pa_n + a^p_{n|p} + R(a_1, \dots, a_{n-1}, b_1, b_2, \dots, b_n) \,.$$

By Theorem 2.12, we conclude that

$$a_n = \frac{1}{p}\left(a_{n|p} - a^p_{n|p} - R(a_1, \dots, a_{n-1}, b_1, b_2, \dots, b_n)\right) \in \mathbb{Z}_p \,,$$

as desired. $\qquad\qquad\qquad\qquad\qquad\qquad\qquad\qquad\qquad\qquad\qquad$ $\square$

Dwork's lemma applies to the function defined by (3.9), with (3.10) showing that the required hypothesis for its application holds. We conclude that $E_p(x) \in 1 + x\mathbb{Z}_p[[x]]$. This is a bit of miracle if we look at the right side of (3.9), which realizes $E_p(x)$ as the exponential of a series whose coefficients diverge p-adically.

One of the results that the alluded miracle for (3.9) encodes is Wilson's theorem. Indeed, the coefficient of x^p in this series is given by

$$\frac{1}{p!} + \frac{1}{p} = \frac{1 + (p-1)!}{p!} \,,$$

and since all the coefficients of the series are in $\mathbb{Z}_p$, among them this particular one, we conclude that $(p-1)! \equiv -1 \bmod p$.

Since $E_p(x) \in \mathbb{Z}_p[[x]]$, the series converges on any disc of radius less than 1. This is precisely the disc of convergence of $E_p(x)$.

3.2 The completion of the algebraic closure of $\mathbb{Q}_p$

By Theorem 2.8, we may conclude readily that $\mathbb{Q}_p$ is not an algebraically closed field. Let $\overline{\mathbb{Q}}_p$ be its algebraic closure, the union of all finite field extensions of $\mathbb{Q}_p$. We see first how to extend the p-adic norm and p-adic order to $\overline{\mathbb{Q}}_p$.

Let K be a finite field extension of degree n of a normed field F, which is assumed to be locally compact. We then have that K is a vector space of dimension n over F, and in such case, we see that norms on K must be equivalent to each other. If K is provided with a field norm, the norm of the power of an element is the said power of the norm of the element. Using this property, we can see easily that all norms on K extending the norm on F must be, in fact, equal to each other. Thus, there exists at most one field norm on K that extends the norm on F.

Let α be an algebraic number of degree n over a field F. We denote by

$$i_\alpha(x) = x^n + a_1 x^{n-1} + \cdots + a_{n-1} x + a_n$$

the unique monic irreducible polynomial with coefficients in F that is annihilated by α. For the finite field extension $K = F(\alpha)$, we define

$$N_{K/F}(\alpha) = (-1)^n a_n \, .$$

This number can be cast also as $\prod_{i=1}^{n} \alpha_i$ where the α_is are all the conjugates of $\alpha_1 = \alpha$ over F, or as the determinant of the automorphism of K given by multiplication by α. In particular, when $K = \mathbb{Q}_p(\alpha)$, we have that $N_{\mathbb{Q}_p(\alpha)/\mathbb{Q}_p}(\alpha) = (-1)^n a_n$, and this is a p-adic number. This discussion provides a way of finding out the proper definition of an extension of the p-adic norm to $\overline{\mathbb{Q}}_p$.

For if K is an extension of F with a norm $\| \ \|$ that extends the norm on F, given $\sigma \in \mathrm{Aut}(K)$, the uniqueness of the norm extension implies that $\|x\|' = \|\sigma(x)\| = \|x\|$. If $K = \mathbb{Q}_p(\alpha)$ is a Galois extension, we can find automorphisms of K taking any conjugate of α to α itself, which implies that all of these conjugates must have the same norm. Since $N_{\mathbb{Q}_p(\alpha)/\mathbb{Q}_p}(\alpha)$ is a p-adic number, we must have that

$$\left\| N_{\mathbb{Q}_p(\alpha)/\mathbb{Q}_p}(\alpha) \right\|_p = \left\| N_{\mathbb{Q}_p(\alpha)/\mathbb{Q}_p}(\alpha) \right\| = \left\| \prod_{i=1}^{n} \alpha_i \right\| = \prod \|\alpha_i\| = \|\alpha\|^n \, ,$$

and so the extended norm of α should be defined by

$$\|\alpha\| = \left\| N_{\mathbb{Q}_p(\alpha)/\mathbb{Q}_p}(\alpha) \right\|_p^{\frac{1}{n}} = \|i_\alpha(0)\|_p^{\frac{1}{n}} = \|a_n\|_p^{\frac{1}{n}} \, .$$

The expression above can be rephrased intrinsically. For if K is any finite field extension of $\mathbb{Q}_p$ that contains α, then $\|\alpha\| = \left\| N_{K/\mathbb{Q}_p}(\alpha) \right\|_p^{\frac{1}{[K:\mathbb{Q}_p]}}$, where $[K:F]$ denotes the degree of the field extension K of F.

Definition 3.8 Let α be an algebraic number over $\mathbb{Q}_p$ contained in a field extension K of $\mathbb{Q}_p$. We define $\|\alpha\|_p$ by

$$\|\alpha\|_p = \left\| N_{K/\mathbb{Q}_p}(\alpha) \right\|_p^{\frac{1}{[K:\mathbb{Q}_p]}}, \tag{3.12}$$

where $N_{K/\mathbb{Q}_p}(\alpha)$ is the value at $x = 0$ of the unique monic irreducible polynomial $i_\alpha(x)$ of α, and $[K : \mathbb{Q}_p]$ is the degree of K over $\mathbb{Q}_p$. $\quad\square$

Expression 3.12 is clearly multiplicative in α, it is only zero when $\alpha = 0$, and it yields the p-adic norm of α in the case when $\alpha \in \mathbb{Q}_p$. Thus, this function is an extension of the p-adic norm function on $\mathbb{Q}_p$. This extension also satisfies the non-Archimedean property (2.2), fact whose proof we leave to the interested reader. Thus, expression 3.12 defines a non-Archimedean norm on $\overline{\mathbb{Q}}_p$ that is an extension of the p-adic norm on $\mathbb{Q}_p$.

For the same reasons as before, the set $\{x \in \overline{\mathbb{Q}}_p : \|x\|_p \leq 1\}$ forms a ring, the ring of integers of $\overline{\mathbb{Q}}_p$.

For any α in a finite field extension K of $\mathbb{Q}_p$ of degree n, we define

$$\mathrm{ord}_p\, \alpha = -\log_p \|\alpha\|_p = -\frac{1}{n} \log_p \left\| N_{K/\mathbb{Q}_p}(\alpha) \right\|_p,$$

where the logarithm is computed in base p. This clearly yields an extension to K of the ord_p function on $\mathbb{Q}_p$. The image of K under this function is an additive subgroup of $\frac{1}{n}\mathbb{Z}$, and therefore, it must be of the form $\frac{1}{e}\mathbb{Z}$ for some positive integer e dividing n. This integer is the *index of ramification* of K over $\mathbb{Q}_p$. If $e = 1$, the field K is said to be an unramified extension of $\mathbb{Q}_p$.

Example 3.9 We can revisit the result in Theorem 2.8 by considering the quadratic equation $f(x) = x^2 - p$. Any root $\alpha \in \overline{\mathbb{Q}}_p$ has p-adic norm $\|\alpha\|_p = 1/p^{\frac{1}{2}}$, and so neither of them is in $\mathbb{Q}_p$. With more generality, if $m/n \in \mathbb{Q}$ with $(m, n) = 1$, then the polynomial $f(x) = x^n - p^m$ is irreducible over $\mathbb{Q}_p$, and each of its roots in $\overline{\mathbb{Q}}_p$ has p-adic norm $1/p^{\frac{m}{n}}$. $\quad\square$

Example 3.10 The primitive p-roots of unity in $\overline{\mathbb{Q}}_p$ are roots of the irreducible polynomial $i_p(x) = x^{p-1} + x^{p-2} + \cdots + x + 1$, and so have degree $p - 1$ over $\mathbb{Q}_p$. If ζ_p is a root of this polynomial, then $\|\zeta_p\|_p = 1$, and any other root is the form ζ_p^k for $1 \leq k \leq p - 1$.

We have that $x i_p(x + 1) = (x + 1)^p - 1$, and the polynomial $i_p(x + 1)$ is irreducible over $\mathbb{Q}_p$, and have degree $p - 1$. Since the roots of $i_p(x + 1)$ have the form $\zeta_p^k - 1$, it follows that $\left\| \zeta_p^k - 1 \right\|_p = \frac{1}{p^{p-1}}$. We have that

$K = \mathbb{Q}_p(\zeta_p)$ is a ramified field extension of $\mathbb{Q}_p$ of degree $p - 1$. Its index of ramification is $p - 1$ also.

In fact, for any $d \geq 1$, the polynomial $i_{p,d}(x) = i_p(x^{p^{d-1}})$ is irreducible over $\mathbb{Q}_p$, and its roots are the p^d-roots of unity in $\overline{\mathbb{Q}}_p$. If ζ_{p^d} is a primitive root, any other is of the form $\zeta_{p^d}^k$ for $1 \leq k \leq p^d - 1$, $p \nmid k$. We have that $\left\| \zeta_{p^d} \right\|_p = 1$ and $\left\| \zeta_{p^d} - 1 \right\|_p = \frac{1}{p^{(p-1)p^{d-1}}}$. Thus, the only p-root of unity in $\mathbb{Q}_p$ is $\xi = 1$ when p is an odd prime, or $\xi = \pm 1$ if $p = 2$.

Let us now consider the situation where the order of the root of unity is not divisible by p. We recall firstly the result in §2.5, where we proved that all $(p - 1)$-roots of unity lie in $\mathbb{Q}_p$. If we consider any primitive $(p^r - 1)$-root of unity $\zeta_{p^r - 1}$, $r \geq 1$, it has degree r over $\mathbb{Q}_p$ and irreducible polynomial $\prod_{0 \leq l \leq r-1}(x - \zeta_{p^r-1}^l)$. Thus, $\left\| \zeta_{p^r-1} \right\|_p = \left\| \zeta_{p^r-1} - 1 \right\|_p = 1$.

Now, if d is any positive rational integer not divisible by p, we let m_d be the smallest positive integer m such that $p^m \equiv 1 \bmod d$. Since the group of roots of the polynomial $x^n - 1$ in $\overline{\mathbb{Q}}_p$ is cyclic (as is every finite multiplicative group in a field), if $\zeta_{p^{m_d}-1}$ is any primitive $(p^{m_d} - 1)$-root of unity, we obtain a d-root of unity ξ by taking

$$\xi = \zeta_{p^{m_d}-1}^{k \frac{p^{m_d}-1}{m_d}},$$

where k is any integer coprime to $(p^{m_d} - 1)/m_d$. We have that $\|\xi\|_p = 1$, $\|\xi - 1\|_p = 1$, and $\xi \in \mathbb{Q}_p$ if, and only if, $m_d = 1$.   □

The ramified notion developed for finite field extensions of $\mathbb{Q}_p$ applies to all elements α in $\overline{\mathbb{Q}}_p$. We say that α is ramified if $\|\alpha\|_p$ is a fractional power p, and define its ramification degree to be the smallest integer $e(\alpha)$ such that $\alpha^{e(\alpha)}$ is unramified. Unramified elements of $\overline{\mathbb{Q}}_p$ are those for which $e(\alpha) = 1$. The example shows that the roots of unity of order coprime to p are unramified elements of $\overline{\mathbb{Q}}_p$.

The union of all finite unramified extensions of $\mathbb{Q}_p$ is denoted by $\mathbb{Q}_p^{ur}$, and it is referred to as the maximal unramified extension of $\mathbb{Q}_p$.

Proposition 3.11   *The maximal unramified extension* $\mathbb{Q}_p^{ur}$ *of* $\mathbb{Q}_p$ *is obtained by adjoining to* $\mathbb{Q}_p$ *all the* n-*roots of unity with* $(n, p) = 1$. *The set of unramified integers* $\mathbb{Z}_p^{ur} = \{x \in \mathbb{Q}_p^{ur} : \|x\|_p \leq 1\}$ *is a (local) ring with a unique maximal ideal* $p\mathbb{Z}_p^{ur}$, *and the residue field* $\mathbb{Z}_p^{ur}/p\mathbb{Z}_p^{ur}$ *is the algebraic closure* $\overline{\mathbb{F}}_p$ *of the field* $\mathbb{F}_p$ *with* p-*elements.*

Proof. The finite unramified extensions of $\mathbb{Q}_p$ are in 1-to-1 correspondence with the finite extensions of $\mathbb{F}_p$. Since the splitting field of $x^{p^k} - x = x(x^{p^k-1} - 1)$ is the unique extension of $\mathbb{F}_p$ of degree k, it follows that $\mathbb{Q}_p$ has a unique unramified extension of degree k for every k, the splitting field of the said polynomial. Thus, a degree k unramified extension of $\mathbb{Q}_p$ is obtained by adjoining the $(p^k - 1)$-roots of unity to $\mathbb{Q}_p$, and $\mathbb{Q}_p^{ur}$, which corresponds to the algebraic closure of $\mathbb{F}_p$, is obtained by adjoining all the $(p^k - 1)$-roots of unity for all k. If k is a multiple of p, the order $m = p^k - 1$ of the root is not divisible by p. On the other hand, for any m coprime to p we have that $p^{\varphi(m)} \equiv 1 \bmod m$, and as we saw in Example 3.10, any m-root of unity is unramified and is obtained as a suitable power of a primitive $(p^{\varphi(m)} - 1)$-root of unity. We therefore conclude that $\mathbb{Q}_p^{ur}$ is obtained by adjoining to $\mathbb{Q}_p$ all the m-roots of unity with $(m, p) = 1$.

Since the p-adic norm is non-Archimedean, the unramified integers $\mathbb{Z}_p^{ur} = \{x \in \mathbb{Q}_p^{ur} : \|x\|_p \leq 1\}$ form a ring. Since every element of $\mathbb{Z}_p^{ur}$ can be written as $p^k u$ for some u with $\|u\|_p = 1$ and some nonnegative integer k, we see that $p\mathbb{Z}_p^{ur}$ is a maximal ideal, and in fact, the only one such. By the discussion above (cf. the arguments in Example 3.10), the residue field $\mathbb{Z}_p^{ur}/p\mathbb{Z}_p^{ur}$ is the algebraic closure $\overline{\mathbb{F}}_p$ of the field $\mathbb{F}_p$ with p-elements. $\qquad\square$

As a metric space, $\overline{\mathbb{Q}}_p$ is not complete. Its completion Ω_p turns out to be algebraically closed.

Let us think of $\mathbb{Q}_p$ as the analogue of $\mathbb{R}$ when in the completion of $\mathbb{Q}$ we use the p-adic norm $\| \ \|_p$ instead of the Euclidean norm. Then Ω_p is to be though as the analogue of $\mathbb{C}$. There is though an interesting p-adic difference: $\mathbb{C}$ is obtained from $\mathbb{R}$ as a field extension of degree two that is algebraically closed and metrically complete, while the algebraic closure $\overline{\mathbb{Q}}_p$ of $\mathbb{Q}_p$ requires an additional metric completion to ultimately yield Ω_p. Neither $\overline{\mathbb{Q}}_p$ nor Ω_p are locally compact: their unit spheres centered at the origin have plenty of sequences without convergent subsequences, for instance, sequences of distinct roots of unity of order coprime to p.

We denote by $\widehat{\mathbb{Z}}_p^{ur} \subset \Omega_p$, the completion of the unramified extension $\mathbb{Z}_p^{ur}$ of $\mathbb{Z}_p$. The theory of arithmetic differential operators we will discuss in §5 over $\mathbb{Z}_p$ has a much richer flavour over $\widehat{\mathbb{Z}}_p^{ur}$. Cf. [8, 6, 15]. The reader will come to appreciate this fact in our outline of the most general version of these operators in §6.2.

Remark 3.12 Any $(p^k - 1)$-root of unity, $k \geq 2$, lies in the boundary of $\widehat{\mathbb{Z}}_p^{ur} \setminus \mathbb{Z}_p$. $\qquad\square$

Remark 3.13 We revisit the statement in Theorem 3.1. The binomial series $\sum_{n \geq 0} \binom{y}{n} x^n$ converges to $(1+x)^y$ for all $x \in \Omega_p$ such that $\mathrm{ord}_p\, x > 1/(p-1)$, and the resulting function is continuous as a function of y. $\qquad \square$

3.3 Zeta functions

Let $\mathbb{F}$ be a field. We denote by $\mathbb{A}_{\mathbb{F}}^n$ the n-dimensional affine space over $\mathbb{F}$, $\mathbb{A}_{\mathbb{F}}^n = \{(x_1, \ldots, x_n) \mid x_i \in \mathbb{F}, 1 \leq i \leq n\}$. the set of n-tuples of elements in $\mathbb{F}$. Then the projective n-space $\mathbb{P}_{\mathbb{F}}^n$ over $\mathbb{F}$ is the set of one-dimensional subspaces of $\mathbb{A}_{\mathbb{F}}^{n+1}$. A point $p \in \mathbb{P}_{\mathbb{F}}^n$ is usually written as a homogeneous vector $[X_0 : \ldots : X_n]$, by which is meant the $\mathbb{F}$-line spanned by $(X_0, \ldots, X_n) \in \mathbb{A}_{\mathbb{F}}^{n+1} \setminus \{0\}$.

The zero locus of a finite family $\{f_i\}_{i \in I}$, $f_i \in \mathbb{F}[X_1, \ldots, X_n]$ defines an affine algebraic variety over $\mathbb{F}$:

$$V_{\{f_i\}_{i \in I}} = \{(X_1, \ldots, X_n) \in \mathbb{A}_{\mathbb{F}}^n : f_i(X_1, \ldots, X_n) = 0\}.$$

The ideal $I(V)$ of V consists of the set of all $\mathbb{F}$-polynomials that vanish on points of V. In fact, any ideal $I \subset \mathbb{F}[X_1, \ldots, X_n]$ defines a corresponding locus of points $V = \{X \in \mathbb{A}_{\mathbb{F}}^n : f(X) = 0 \text{ for all } f \in I\}$, and the Hilbert's basis theorem asserts that I has a finite basis $f_1, \ldots, f_k$ so that this locus is in effect an affine algebraic variety. However, the corresponding ideal $I(V)$ could be larger than I.

A polynomial $f \in \mathbb{F}[X_0, \ldots, X_n]$ does not in general descend to a function on $\mathbb{P}_{\mathbb{F}}^n$. However, if f is a homogeneous polynomial of degree d, the notion of zeroes of f in $\mathbb{P}_{\mathbb{F}}^n$ makes sense because we have the relation $f(\lambda X_0, \ldots, \lambda X_n) = \lambda^d f(X_0, \ldots, X_n)$. A projective algebraic variety $V \subset \mathbb{P}_{\mathbb{F}}^n$ over $\mathbb{F}$ is defined to be the zero locus of a collection of homogeneous polynomials in $\mathbb{F}[X_0, \ldots, X_n]$, and its ideal $I(X)$ consists of the set of polynomials that vanish on X.

For instance, the *rational normal curve* $C_d \subset \mathbb{P}_{\mathbb{F}}^n$ of degree d is defined to be the image of the map $\mathbb{P}_{\mathbb{F}}^1 \to \mathbb{P}_{\mathbb{F}}^d$ given by

$$[X_0 : X_1] \mapsto [X_0^d : X_0^{d-1} X_1 : \ldots : X_0 X_1^{d-1} : X_1^d] = [Z_0 : \ldots : Z_d].$$

It is the common zero locus of the polynomials $p_{ij} = Z_i Z_j - Z_{i-1} Z_{j+1}$ for $1 \leq i \leq j \leq d-1$, and has associated with it the ideal $I(C_d) := \{f \in \mathbb{F}[Z_0, \ldots, Z_n] \mid f \equiv 0 \text{ on } C_d\}$. This ideal is, in fact, generated by the family of polynomials p_{ij}.

If $\mathbb{E}$ is a field extension of $\mathbb{F}$ and $V = V_{\{f_i\}_{i \in I}}$ is a projective algebraic variety over $\mathbb{F}$ defined by homogeneous polynomials $f_i \in \mathbb{F}[X]$, then V is defined also over the extension $\mathbb{E}$ as the coefficients of the f_is whose zeros define V over $\mathbb{F}$ are also all elements of $\mathbb{E}$. Thus, we can talk about the set $V(\mathbb{E})$ of $\mathbb{E}$-points of V, $V(\mathbb{E}) = \{(X_0, \ldots, X_n) \in \mathbb{P}^n_{\mathbb{E}} : f_i(X_0, \ldots, X_n) = 0, \ i \in I\}$. This makes of the $\mathbb{F}$-variety V a functor from the category of field extensions of $\mathbb{F}$ and their morphisms to a suitable category of sets and morphisms, with the functor mapping an extension $\mathbb{E}$ of $\mathbb{F}$ to the set $V(\mathbb{E})$ of $\mathbb{E}$-points of the variety. Let us observe in passing that using *restrictions* when possible, we may also carry out this idea in the opposite direction, and find the points of a variety that lie on a subring of $\mathbb{F}$ when the variety in question is defined by polynomials whose coefficients are elements of the subring (see [31] for an elementary expansion on this point that relates to the roots of unity).

Let us assume now that $\mathbb{F}$ is a finite field $\mathbb{F}_q$ of characteristic p. A simple combinatorial argument shows that number of bases of the vector space $\mathbb{F}_q^n$ over $\mathbb{F}_q$ is given by

$$(q^n - 1)(q^n - q) \cdots (q^n - q^{n-1}) = q^{\frac{n(n-1)}{2}} (q - 1)^n [n]_q!,$$

where

$$[n]_q = 1 + q + q^2 + \cdots + q^{n-1},$$

and where the factorial is defined by

$$[n]_q! = [1]_q \cdot [2]_q \cdots [n]_q.$$

Proceeding similarly, we see that the number of linearly independent k-element subsets of $\mathbb{F}_q^n$ is equal to

$$(q^n - 1)(q^n - q) \cdots (q^n - q^{k-1}) = q^{\frac{k(k-1)}{2}} (q - 1)^k [n]_q!/[n - k]_q!,$$

and so if $k \leq n$, the number of subspaces of $\mathbb{F}_q^n$ of dimension k is

$$\binom{n}{k}_q = \frac{[n]_q!}{[n - k]_q![k]_q}.$$

In particular, we see that the number of points in $\mathbb{P}^n_{\mathbb{F}_q}$ is equal to $q^n + q^{n-1} + \cdots + q + 1$.

Given a variety V defined over $\mathbb{F}_q$, we denote by N_j the number of $\mathbb{F}_{q^j}$-points of V. We may encode the combinatorial information given by all of the numbers N_j, $j = 1, 2, \ldots$ in the Weil zeta function of V that is defined as

$$\zeta(V/\mathbb{F}_q, t) = \exp\left(\sum_{j=1}^{\infty} \frac{N_j}{j} t^j\right) \in 1 + t\mathbb{Q}[[t]], \qquad (3.13)$$

Example 3.14   We consider the case where $V = \mathbb{P}^n_{\mathbb{F}_q}$. Then the number N_j of $\mathbb{F}_{q^j}$-points of this variety is $q^{jn} + q^{j(n-1)} + \cdots + q^j + 1$, and we have that

$$\zeta(\mathbb{P}^n/\mathbb{F}_q, t) = \frac{1}{(1-t)\cdots(1-q^n t)} = \prod_{j=0}^{n} \frac{1}{1 - q^j t}.$$

This is a rational function with coefficients in $\mathbb{Z}$ that satisfies the identity $Z(\mathbb{P}^n/\mathbb{F}_q, 1/q^n t) = q^{\frac{n(n+1)}{2}} t^{n+1} Z(\mathbb{P}^n/\mathbb{F}_q, t)$. Let us think of the integer n as a topological invariant of V. Then this last relationship illustrates the close interconnection between topological invariants of V and the geometry of the algebraic variety over a finite field.

If we consider the case where $V = \mathbb{A}^n_{\mathbb{F}_q}$ instead, then $N_j = q^{jn}$, and we have that

$$\zeta(\mathbb{A}^n/\mathbb{F}_q, t) = \frac{1}{1 - q^n t}.$$

$\square$

Remark 3.15   We may also define the zeta function of nonnegative elliptic pseudodifferential operators. Their connection with the number theoretic considerations introduced above is made by the spectral analysis of linear operators on Banach spaces. We illustrate quickly this issue via the case of the Neumann operator.

Let us consider any manifold M that is the the the boundary of a closed connected Riemannian manifold with boundary $\tilde{M}$. The Riemannian structure allows us to define a vector field that is normal to the boundary, the normal vector field. Given a distribution u_0 on M, let u be the unique solution to the Dirichlet problem on $\tilde{M}$ that has boundary value u_0. The Neumann operator N acts on u_0, and maps it to the restriction to the boundary of the normal derivative of u.

The Neumann operator N is a nonnegative elliptic pseudodifferential operator. As a function on the cotangent bundle T^*M, its principal symbol is $\sigma(N)(y, \eta) = i|\eta|$ [46]. The kernel of N is the set of constant functions.

The manifold M itself is Riemannian with Riemannian structure induced by the Riemannian structure on $\tilde{M}$. Let Δ be the Laplacian of the metric. Then $\sqrt{\Delta} = N$ modulo terms of order zero or less, and in this sense, N is a "scalar" Dirac operator. We consider the self-adjoint extension of N to $L^2(M)$, which we shall denote by N also.

Let T be any nonnegative self-adjoint operator on a Hilbert space with discrete spectrum. Let $0 < \lambda_1 \le \lambda_2 \le \cdots$ be the sequence of nonzero eigenvalues of T counted with multiplicity. The zeta function of T is defined by

$$\zeta_T(s) = \sum_{n=1}^{\infty} \frac{1}{\lambda_n^s}.$$

This function is considered with domain the region of the complex plane where the series converges.

The determinant of T is defined in terms of its zeta function by the identity $\det T = e^{-\zeta_T'(0)}$.

Let $N = N_{\mathbb{S}_r^1}$ be the Neumann operator on the circle of radius r in $\mathbb{C}$ centered at the origin (see [45] for the key role this operator plays in the analysis of the Zaremba problem on planar domains). Then we have that

$$\zeta_N(s) = 2r^s \sum_{n=1}^{\infty} \frac{1}{n^s} = 2r^s \zeta(s).$$

where $\zeta(s)$ is the Riemann zeta function. Indeed, we just need to observe that

$$\frac{1}{\sqrt{2\pi r}} e^{in\theta/r}$$

is an eigenfunction of N of eigenvalue $|n|/r$, and the stated assertion follows easily from that.

Using this relationship and the fact that $\zeta(0) = -1/2$ and $\zeta'(0) = -(1/2)\log 2\pi$, it follows $\zeta_N'(0) = -\log(2\pi r)$. Thus, the determinant of N over the r-circle is given by

$$\det N_{\mathbb{S}_r^1} = e^{-\zeta_N'(0)} = 2\pi r \,,$$

the length of $\mathbb{S}_r^1$. □

It would be remarkable to produce an interpretation of the zeta function of nonnegative elliptic pseudodifferential operators in terms of the arithmetic differential operators, which are the subject of our work here. At present, we are far from being able to do so.

We call a variety V defined over $\mathbb{F}$ a hypersurface if V is defined as the zero locus of a single polynomial in $\mathbb{F}[x]$.

Lemma 3.16 *Let V be a variety over the finite field $\mathbb{F}_q$. Then its zeta function $\zeta(V/\mathbb{F}_q, t)$ in (3.13) is the product of the zeta functions of a finite number of hypersurfaces, or their inverses, and in fact, $\zeta(V/\mathbb{F}_q, t) \in 1 + t\mathbb{Z}[[t]]$.*

Proof. By the Hilbert's basis theorem, the ideal $I(V)$ of V has a finite basis $\{f_1, \ldots, f_k\}$, and we have $V = V_{\{f_1, \ldots, f_k\}}$. Given a finite number of polynomials $f_{i_1}, \ldots, f_{i_l}$, we denote by $H_{f_{i_1}, \ldots, f_{i_l}}$ the hypersurface defined by the product $f_{i_1} \cdots f_{i_l}$, and by $N_{j, H_{f_{i_1}, \ldots, f_{i_l}}}$ its number of $\mathbb{F}_{q^j}$-points. By the inclusion-exclusion combinatorial principle, we have

$$N_j = \sum_{i_1 < \cdots < i_l} (-1)^{l+1} N_{j, H_{f_{i_1}, \ldots, f_{i_l}}},$$

and so

$$\zeta(V/\mathbb{F}_q, t) = \prod_{i_1 < \cdots < i_l} \zeta(H_{f_{i_1}, \ldots, f_{i_l}} / \mathbb{F}_q, t)^{(-1)^{l+1}},$$

which proves the first assertion.

Let us now consider a hypersurface H_f. Let E be an extension of $\mathbb{F}_q$, and consider an E-point p. We let r_0 be the smallest positive integer such that all of the components of p are in $\mathbb{F}_{q^{r_0}}$. Each component of p has r_0 conjugates, and by varying all of these components among their conjugates, we form a set of r_0 points $p_1, \ldots, p_{r_0}$. These points must be distinct. They are $\mathbb{F}_{q^r}$-points of H_f if, and only if, r_0 divides r, so they contribute r_0 to each of the numbers N_{jr_0, H_f}, $j = 1, 2, \ldots$, and their contribution to $\zeta(H_f/\mathbb{F}_q, t)$ is given by $\exp\left(\sum_{j=1}^{\infty} r_0 t^{jr_0}/jr_0\right) = (1 - t^{r_0})^{-1} = \sum_{j=0}^{\infty} t^{jr_0} \in 1 + t\mathbb{Z}[[t]]$. It follows that the zeta function of H_f must be in $1 + t\mathbb{Z}[[t]]$, as this function is the product of series of this type. $\square$

Example 3.17 In the analysis of the Weil function of a variety, the lemma above gives a special importance to the hypersurface case. We discuss here the zeta function of hypersurfaces when $n = 1$, hypersurfaces defined by polynomials of one variable. We will do so assuming that $q = p$, so V is defined by a polynomial $f(x)$ in $\mathbb{F}_p[x]$. We shall assume also that $f(x)$ does not have multiples roots in $\overline{\mathbb{F}}_p$.

Let us decompose the defining polynomial $f(x)$ into a product of irreducible factors $f = f_1 \cdots f_k$, and let d_i denote the degree of f_i, so $\sum_i d_i = d$, where d is the degree of f. We then have that $N_{j, H_{f_i}} = d_i$ if $d_i \mid j$ and it is zero otherwise. For each root of f_i induces a homomorphism $\mathbb{F}_p[x]/(f_i) \to \overline{\mathbb{F}}_p$ whose image is $\mathbb{F}_{p^{d_i}}$, and therefore, the set of all roots of f_i must be $\mathbb{F}_{p^{d_i}}$. Now if f_i has a root in $\mathbb{F}_{p^k}$, then $\mathbb{F}_{p^{d_i}} \subset \mathbb{F}_{p^k}$

and $\mathbb{F}_{p^k}$ is a field extension of $\mathbb{F}_{p^{d_i}}$. Then $d_i \mid k$ and all the roots of f_i are in $\mathbb{F}_{p^k}$. Conversely, if $d_i \mid k$ then $\mathbb{F}_{p^{d_i}} \subset \mathbb{F}_{p^k}$, and the larger field must contain all the roots of f_i, which finishes the proof of the assertion.

Therefore, we have that $\zeta(H_{f_i}/\mathbb{F}_p, t) = 1/(1 - t^{d_i})$, and

$$\zeta(V_f/\mathbb{F}_p, t) = \prod_{i=1}^{k} \zeta(H_{f_i}/\mathbb{F}_p, t) = \prod_{i=1}^{k}(1 - t^{d_i})^{-1}.$$

Thus, $\zeta(V_f/\mathbb{F}_p, t)$ is a rational function with coefficients in $\mathbb{Z}$. Further, $\zeta(V_f/\mathbb{F}_p, t)$ is a function of the form $1/p(t)$, where $p(t)$ can be written as $\prod_{i=1}^{d}(1 - \zeta_i t) \in \mathbb{Z}[t]$, $\mid \zeta_i \mid = 1$, $i = 1, \ldots, d$. Finally, $\zeta(V_f/\mathbb{F}_p, 1/t) = (-1)^k t^d \zeta(V_f//\mathbb{F}_p, t)$.

Encoded in the zeta function $\zeta(V_f//\mathbb{F}_p, t)$ we have all the information concerning the field extensions $\mathbb{F}_q$ where all the roots of the polynomial f lie. □

The zeta function (3.13) of a smooth projective variety V defined over a finite field $\mathbb{F}_q$ was the subject of several famous conjectures by A. Weil in 1949 [48]. The conjectures state that the zeta function of V has properties similar to those of the zeta function of the elementary Example 3.17 analyzed above. They are:

1. $\zeta(V/\mathbb{F}_q, t)$ is a rational function with coefficients in $\mathbb{Z}$.
2. The topology of the variety and its zeta function are intimately related in that

$$\zeta(V/\mathbb{F}_q, t) = \frac{p_1(t)p_3(t)\cdots p_{2n-1}(t)}{p_0(t)p_2(t)\cdots p_{2n}(t)},$$

where n is the dimension of V, $p_k(t) \in \mathbb{Z}[t]$, $p_0(t) = (1 - t)$, $p_k(t) = \prod_{j=1}^{b_k}(1 - \alpha_{k,j}t)$, for some $\alpha_{k,j} \in \mathbb{C}$ such that $\mid \alpha_{k,j} \mid = q^{\frac{j}{2}}$, $1 \leq k \leq 2n - 1$, and $p_{2n}(t) = (1 - q^n t)$. When V is the reduction mod p of a smooth variety over a field embedded in $\mathbb{C}$, the degrees b_k of the polynomials p_k can be interpreted as the Betti numbers of the complex points of V.
3. If E is the Euler characteristic of V, then

$$\zeta(V/\mathbb{F}_q, 1/q^n t) = \pm q^{\frac{nE}{2}} t^E \zeta(V/\mathbb{F}_q, t),$$

relation that implies that the sets $\{\alpha_{2n-i,l}\}_l$ and $\{q^n/\alpha_{i,l}\}_l$ coincide.

The first major step in the settling of these conjectures was taken by Dwork in 1960 [23, 24]. He proved the rationality of the zeta function by applying techniques from p-adic analysis to the study of the problem. He produced suitable lifts of characters on $\mathbb{F}_q$ to Ω_p, space where the said techniques could be then used. We now discuss this issue further because of its relation to lifts of Frobenius mappings, which do play a central role in the theory of arithmetic differential equations. Dwork's use of lifts of Frobenius is inspired in a philosophy that is perpendicular to ours. Indeed, he views the p-adic numbers as arguments of functions, and not as functions themselves. We shall not use Dwork's idea otherwise, and our discussion of his theorem will be very sketchy. We refer the reader to Chapter V in [29] or II.6 and II.7 in [25] for an exposition of this in its entirety. We shall follow here the first of these references somewhat.

By Lemma 3.16, it suffices to carry out the proof for a nonsingular projective hypersurface V defined over $\mathbb{F}_q$. In this case, if $f(X_0, \ldots, X_n)$ is the defining polynomial of V, the nonsingular condition on V means that f and $(\partial_{X_0} f, \ldots, \partial_{X_n} f)$ do not have common zeroes in $\overline{\mathbb{F}}_q$, but in effect, Dwork's result is general, and it does not require this condition.

The dimension of the hypersurface V is $n - 1$. Using the cell decomposition $\mathbb{P}_{\mathbb{F}}^n = (\mathbb{P}_{\mathbb{F}}^n \setminus \mathbb{P}_{\mathbb{F}}^{n-1}) \cup (\mathbb{P}_{\mathbb{F}}^{n-1} \setminus \mathbb{P}_{\mathbb{F}}^{n-2}) \cup \cdots \cup (\mathbb{P}_{\mathbb{F}}^1 \setminus \mathbb{P}_{\mathbb{F}}^0) \cup \mathbb{P}_{\mathbb{F}}^0$, we can express V as the disjoint union of affine hypersurfaces of different dimensions, and so the rationality of $\zeta(V/\mathbb{F}_q, t)$ will follow if we can prove the corresponding statement in the case where V is an affine hypersurface defined by $f \in \mathbb{F}_q[x]$, which we assume hereupon. The argument is by induction on n, the statement being clear when $n = 1$.

The bulk of the work lies in the proof that the zeta function of an affine hypersurface V is a meromorphic function. In order to do that, let us consider the modified zeta function

$$\zeta'(V/\mathbb{F}_q, t) = \exp\left(\sum_{j=1}^{\infty} \frac{N_j'}{j} t^j\right), \tag{3.14}$$

where N_j' is now the number of $\mathbb{F}_{q^j}$-points $(x_1, \ldots, x_n)$ of V with no zero coordinate. Then $\zeta(V/\mathbb{F}_q, t) = \zeta'(V/\mathbb{F}_q, t) \exp\left(\sum_j (N_j - N_j') t^j / j\right)$. If H_i is the zero locus of $g_i = f(x_1, \ldots, x_{i-1}, 0, x_{i+1}, \ldots, x_n)$, the exponential factor is the zeta function of $\cup_{i=1}^n H_i$. Each H_i is either n-affine space or an affine variety of dimension less or equal than $n - 2$, and their zeta functions are known either by using Example 3.14 or by the induction hypothesis, and regardless of the case, they have the desired properties. By an application of the inclusion-exclusion principle, cf. with the proof of Lemma 3.16, we see that the said exponential is an element of $1 + t\mathbb{Z}[[t]]$, and by Lemma 3.16, we conclude that $\zeta'(V/\mathbb{F}_q, t) \in 1 + t\mathbb{Z}[[t]]$ also.

Since the modified zeta function $\zeta'(V/\mathbb{F}_q, t)$ in (3.14) is an element of $1 + t\mathbb{Z}[[t]]$, we may try to lift it to a p-adic meromorphic function on Ω_p. Since the multiplicative group $\mathbb{F}_q^\times$ of nonzero elements of $\mathbb{F}_q$ is cyclic of order $q - 1$, the integer N_j' can also be described as the number of $\mathbb{F}_q$-points $(x_1, \ldots, x_n)$ of V such that $x_i^{q^j} = 1$, $i = 1, \ldots, n$. This is the key reason for the introduction of $\zeta'(V/\mathbb{F}_q, t)$, for now we can use this condition in order to count N_j' through a function that have a suitable meromorphic lift to Ω_p. This will show that $\zeta(V/\mathbb{F}_q, t)$ itself is meromorphic. Then it can be proven that $\zeta(V/\mathbb{F}_q, t)$ is actually rational.

Given a finite field extension E of $\mathbb{F}_q$ of degree j, we have the trace $\mathrm{Tr}_{E/\mathbb{F}_q}$ and norm $N_{E/\mathbb{F}_q}$ operations. These are defined by $\mathrm{Tr}_{E/\mathbb{F}_q}(a) = a + a^q + a^{q^2} + \cdots + a^{q^{j-1}}$ and $N_{E/\mathbb{F}_q}(a) = a \cdot a^q \cdots a^{q^{j-1}}$, the sum and product of the conjugates of a, respectively. Modulo a rational number, these two operations can be cast as the trace and determinant of the mapping $E \ni b \to ab \in E$ (cf. with the use of the same mapping in §3.2 when extending the p-adic norm to $\overline{\mathbb{Q}}_p$). They define surjective homomorphism $\mathrm{Tr}_{E/\mathbb{F}_q} : E \to \mathbb{F}_q$ and $N_{E/\mathbb{F}_q} : E^\times \to \mathbb{F}_q^\times$. We use the first of these to count N_j'.

Let $\xi \in \Omega_p$ be a nontrivial p-root of unity in Ω_p. Then

$$
\begin{array}{ccc}
\mathbb{F}_q & \to & \Omega_p \\
a & \mapsto & \xi^{\mathrm{Tr}_{\mathbb{F}_q/\mathbb{F}_p}(a)}
\end{array}
\tag{3.15}
$$

is an Ω_p-character on the additive group $\mathbb{F}_q$, and we have that

$$
\sum_{a \in \mathbb{F}_q} \xi^{\mathrm{Tr}_{\mathbb{F}_q/\mathbb{F}_p}(ba)} = \left\{ \begin{array}{ll} q & \text{if } b = 0, \\ 0 & \text{if } b \in \mathbb{F}_q^\times. \end{array} \right.
$$

By this identity, we see that each $\mathbb{F}_{q^j}$-point $(x_1, \ldots, x_n)$ in V, all of whose components are nonzero, contributes q^j to $\sum_{x_0 \in \mathbb{F}_{q^j}} \xi^{x_0 f(x_1, \ldots, x_n)}$, and so taking away the contributions from $x_0 = 0$, we obtain that

$$
\sum_{x_0, \ldots, x_n \in \mathbb{F}_{q^j}^\times} \xi^{\mathrm{Tr}(x_0 f(x_1, \ldots, x_n))} = q^j N_j' - (q^j - 1)^n.
\tag{3.16}
$$

At this point, we must analyze the left side of this expression in detail.

Given $a \in \mathbb{F}_q$, its Teichmüller representative $t = t_a \in \widehat{\mathbb{Z}}_p^{ur} \subset \Omega_p$ is defined to be a root of $x^q - x$ such that $t \equiv a \bmod p$. By Proposition 3.11, the set $\{t_a\}_{a \in \mathbb{F}_q}$ is contained in the splitting field of $x^q - x$ over $\mathbb{Q}_p$. (The reader may look back at our discussion in §2.5, where we proved explicitly that for any element $\mathbb{F}_p \ni a \neq 0$, there was a $(p-1)$-root of unity α_a such that $\alpha_a \equiv a \bmod p$; cf. with the discussion of Teichmüller representatives in Example 4.8.) We now have the following key result due to Dwork (see Chapter V, §2 of [29] for its proof):

Lemma 3.18 *Suppose that* $q = p^j$ *for some positive integer* j. *There exists a p-adic power series* $\Theta(x) = \sum a_n x^n \in \mathbb{Q}_p(\xi)[[x]]$ *with* $\operatorname{ord}_p a_n \geq n/(p-1)$ *such that the character function* (3.15) *on* $\mathbb{F}_q$ *can be obtained by evaluating* $\Theta(x)\Theta(x^p)\cdots\Theta(x^{p^{j-1}})$ *at the Teichmüller lift* t *of* a.

This result is rather subtle. For instance, since the Teichmüller lift α_a of $a \in \mathbb{F}_p$ is such that $\alpha_a \equiv a \bmod p$, and since the power in $\mathbb{Z}_p$ of the p-root of unity ξ depends only on its congruence class mod p, we have $\xi^{\operatorname{Tr} a} = \xi^{\operatorname{Tr} \alpha_a}$. However, as explained in Example 3.10 (cf. remark 3.12), a $(p^l - 1)$-root of unity lies in the boundary of $\widehat{\mathbb{Z}_p^{ur}} \subset \Omega_p$, while, for instance, the Artin–Hasse exponential function has radius of convergence strictly less than 1, and could not include such a point in its domain of definition. A refined argument must be used in order to yield a function that lifts $a \mapsto \xi^{\operatorname{Tr} a}$ to a function on a disc in Ω_p whose radius is greater or equal than 1. Once that is accomplished, the general techniques of p-adic analysis can then be applied to study the lifted function in Ω_p.

The function $\Theta(x)$ used by Dwork is a suitable modification of the Artin–Hasse function $E_p(x)$. It can be described in simple terms best by using the binomial functions of Theorem 3.1. Let $\lambda = \xi - 1$. As we saw in Example 3.10, λ has p-adic order $1/(p-1)$, and if we view $g(x) = g(x, \lambda) = (1 + \lambda)^x$ as a power series in x, the series will not converge at the xs of interest. So $g(x)$ is modified to

$$g(x, y) = (1 + y)^x (1 + y^p)^{\frac{x^p - x}{p}} (1 + y^{p^2})^{\frac{x^{p^2} - x^p}{p^2}} \cdots$$

and Θ is defined to be

$$\Theta(x) = g(x, \lambda) = \sum_{n=0}^{\infty} a_n x^n.$$

We can then verify the growth in the order of the a_ns that is stated in the Lemma. This estimate for the p-adic order ensures that this series converges on the disc $\{x \in \Omega_p : \|x\|_p < p^{\frac{1}{p-1}}\}$.

The root of unity ξ corresponds to a unique solution π of the equation $x^{p-1} = -p$ in Ω_p. Then we can define

$$E_\pi(x) = e^{\pi(x - x^p)} = e^{\left(\pi x + \frac{(\pi x)^p}{p}\right)} = E_p(\pi x) \prod_{j=2}^{\infty} e^{-\frac{(\pi x)^{p^j}}{p^j}}.$$

This function is equal to $\Theta(x)$. Indeed, $E_\pi(1)$ is the unique p-root of unity congruent to $1 + \pi$ modulo π^2, and for any $c \in \mathbb{Z}_p$ such that $c^p = c$, we have that $E_\pi(c) = E_\pi(1)^c$, so if $x^{p^j} = x$, then $E_\pi(1)^{\sum_{l=0}^{j-1} x^{p^l}} = \Theta(x)\Theta(x^p) \cdots \Theta(x^{p^{j-1}})$. The evaluation of $\Theta(x)$ for any x such that $\|x\|_p \le 1$ is achieved by substitution into the defining power series, and in general it cannot be obtained directly by substitution into $e^{\pi(x - x^p)} = e^{\left(\pi x + \frac{(\pi x)^p}{p}\right)}$.

Let us return to (3.16), and complete the promised sketch of the proof. The coefficients of $x_0 f(x_1, \ldots, x_n)$ are all elements of $\mathbb{F}_q$, so they can be lifted to their Teichmüller representatives in Ω_p to obtain a polynomial function $F(X) = F(X_0, \ldots, X_n) = \sum_{l=1}^N a_l X^{I_l} \in \Omega_p[X_0, \ldots, X_n]$. Here $I_l = (i_{l0}, i_{l1}, \ldots, i_{ln})$ is a tuple of integers, and X^{I_l} has the usual meaning. Then using Lemma 3.18 in (3.16), we obtain that

$$q^j N_j' = (q^j - 1)^n + \sum_{\substack{x_0, \ldots, x_n \in \Omega_p \\ x_i^{q^j - 1} = 1, \ i = 0, \ldots n}} \prod_{l=1}^n \Theta(a_l x^{I_l})\Theta(a_l^p x^{pI_l}) \cdots \Theta(a_l^{p^{j-1}} x^{p^{j-1} I_l}).$$

Let $G(X_0, \ldots, X_n) = \Theta(a_l x^{I_l})\Theta(a_l^p x^{pI_l}) \cdots \Theta(a_l^{p^j} x^{p^{j-1} I_l})$, and let T_q be the operator acting on series and defined by $T_q(\sum a_I X^I) = \sum a_{qI} X^I$, where qI is the tuple obtained by multiplying each entry of I by q. Since the a_ls are in the unit disc in Ω_p, the coefficients of $G(x)$ are such that the traces of any power of the linear operator $\Psi = T_q \circ G$ are well defined (see [29, Chapter V §3]), and we obtain that

$$q^j N_j' = (q^j - 1)^n + (q^j - 1)^{n+1} \mathrm{Tr}(\Psi^j),$$

so

$$N_j' = \sum_{l=0}^n (-1)^l \binom{n}{i} q^{j(n-l-1)} + \sum_{l=0}^{n+1} (-1)^l \binom{n+1}{l} q^{j(n-l)} \mathrm{Tr}(\Psi^j).$$

Let $\Delta(t) = \exp(-\sum_{j=1}^\infty \mathrm{Tr}(\Psi^j) t^j / j)$, the exponential computed in Ω_p. This is a meromorphic function of t [29, Chapter V, §3], and by the identity above, we obtain that

$$\zeta'(H_f / \mathbb{F}_q, t) = \prod_{l=1}^n (1 - q^{n-l-1} t)^{(-1)^{l+1} \binom{n}{l}} \prod_{l=0}^{n+1} \Delta(q^{n-l} t)^{(-1)^{l+1} \binom{n+1}{l}},$$

is also a meromorphic function.

Once it is known that $\zeta'(H_f / \mathbb{F}_q, t)$ is meromorphic, it can be proved that this meromorphic function is in fact a rational function [29, Chapter V, §5]. Thus, the zeta function of H_f itself is rational.

4

Analytic functions on $\mathbb{Z}_p$

Let us pause to introduce some terminology before proceeding any further with our work. We will need to consider some power series in several variables below, and so we give the next definition with a higher degree of generality from the one we have used until now.

Given a multi-index $\alpha = (\alpha_0, \ldots, \alpha_k)$ of nonnegative rational integers, we shall say that $\alpha \geq 0$, and use the conventional expression x^α to denote the monomial $x_0^{\alpha_0} \cdots x_k^{\alpha_k}$. By the weight $|\alpha|$ of the multi-index α, we mean the rational integer $|\alpha| = \alpha_0 + \cdots + \alpha_k$.

Definition 4.1 We say that $F(x) = \sum_{\alpha \geq 0} a_\alpha x^\alpha \in \mathbb{Q}_p[[x_0, \ldots, x_k]]$ is a restricted power series if $\lim_{|\alpha| \to \infty} a_\alpha = 0$. $\qquad\square$

Let

$$F(x) = \sum_{j=0}^{\infty} a_j x^j \tag{4.1}$$

be an element of $\mathbb{Q}_p[[x]]$. For this series to converge on $\mathbb{Z}_p$, it is both necessary and sufficient that

$$\lim_{j \to \infty} a_j = 0 \quad \text{in} \quad \mathbb{Q}_p. \tag{4.2}$$

In this case, the series $F(x)$ defines an element of $C(\mathbb{Z}_p, \mathbb{Q}_p)$. Further, as we indicated earlier, the series of n term-by-term differentiations

$$F^{(n)}(x) = \sum_{j=n}^{\infty} j(j-1) \cdots (j-n+1) a_j x^{j-n},$$

is also a convergent series on $\mathbb{Z}_p$, and it defines a continuous function. Thus, power series of the type (4.1) whose coefficients satisfy (4.2) are closed under differentiation, and that makes the theory of analytic functions over $\mathbb{Q}_p$ simpler than the corresponding theory over $\mathbb{C}$.

Definition 4.2 A function $f : \mathbb{Z}_p \to \mathbb{Q}_p$ is said to be *analytic* if there exists an integer $N \geq 1$ and power series $F_0, F_1, \ldots, F_{p^N-1} \in \mathbb{Q}_p[[x]]$ with radius of convergence $1/p^N$, such that

$$f(x) = F_k(x - k)$$

for each $k \in \{0, 1, \ldots, p^N - 1\}$, and each $x \in k + p^N\mathbb{Z}_p$.   □

Remark 4.3 An analytic function f on $\mathbb{Z}_p$ defines an element of $C(\mathbb{Z}_p, \mathbb{Q}_p)$. By compactness of $\mathbb{Z}_p$, all elements of $C(\mathbb{Z}_p, \mathbb{Q}_p)$ are bounded. If f is locally given by a series of the type (4.1) with coefficients satisfying (4.2), then the coefficients of the series must be bounded. Thus, for a suitable integer n, the function $p^n f$ ranges in $\mathbb{Z}_p$, that is to say, it is an element of $C(\mathbb{Z}_p, \mathbb{Z}_p)$, and can be represented locally by series as in Definition 4.2 whose coefficients are in $\mathbb{Z}_p$.   □

It is natural to ask for a characterization of the functions in $C(\mathbb{Z}_p, \mathbb{Z}_p)$ that are analytic. We put this task in some perspective by reconsidering the Mahler's expansion of a function $f(x)$ in this space, whose coefficients a_n form a restricted sequence in $\mathbb{Z}_p$. By Mahler's expansion, it follows immediately that $f(x)$ is the uniform limit of polynomials, generalizing in this sense Weierstrass' theorem. The relationship (3.4) between the a_ns and the values of f over the rational integers has another important consequence: the sequence of partial sums of the Mahler's expansion of $f(x)$ yields an optimal polynomial of degree n approximating f uniformly. These approximating optimal polynomials are not unique, but it is important to have one produced in a canonical manner, and in this case, the polynomial is produced out of the knowledge of the values of f on the set $\{0, \ldots, n\}$.

However, it is difficult to find conditions on the growth of the $\|a_n\|_p$s that would ensure that $f'(x)$ exists and is a continuous function also. Mahler shows [36, ch. 7, Theorem 4] that if $\lim_{n\to\infty} n \|a_n\|_p = 0$, then $f(x) \in C(\mathbb{Z}_p, \mathbb{Z}_p)$ has a continuous derivative. In general, if a function has a Mahler series with coefficients a_n and its derivative exists and is continuous in $\mathbb{Z}_p$, then the Mahler's coefficients of the derivative are given by $a'_n = \sum_{k=1}^{\infty} (-1)^{k-1} a_{k+n}/n$, which therefore, must be a null sequence. But the converse does not hold.

Further, as in the real case, there are continuous functions on $\mathbb{Z}_p$ that have continuous derivatives of all order, but are not analytic as defined previously [36]. Indeed, the function

$$f(x) = \sum_{n=0}^{\infty} a_n \begin{pmatrix} x \\ n \end{pmatrix} \qquad a_n = p^{[\sqrt{n}]},$$

is an element of $C(\mathbb{Z}_p, \mathbb{Q}_p)$ and can be continuously differentiated infinitely many times. This follows [36, ch. 7, Theorem 5] by the fact that for all $\beta > 0$ we have that $\lim_{n \to \infty} n^{\beta} \|a_n\|_p = 0$. However, for $f(x)$ to be analytic, it must be the case that $\|a_n\|_p < p^{-n/p}$ [36, p. 85], and the coefficients $p^{[\sqrt{n}]}$ do not satisfy that condition.

Characterizing p-adic analytic functions becomes a rather difficult task.

Definition 4.4 A function $f : \mathbb{Z}_p \to \mathbb{Z}_p$ is said to be *analytic of level m*, if for any $a \in \mathbb{Z}_p$ there exists a restricted power series $F_a \in \mathbb{Z}_p[[x]]$ such that

$$f(a + p^m u) = F_a(u)$$

for all $u \in \mathbb{Z}_p$. We say that the collection of series F_a *represents f*. $\square$

The definition above is not conventional. It was used in [14] for reasons that will become clear later on.

Remark 4.5 We have the following:

1. In Definition 4.4, it is enough to take the as in a complete residue system mod p^m in $\mathbb{Z}_p$.
2. If we consider functions in $C(\mathbb{Z}_p, \mathbb{Z}_p)$, the notions of analyticity in Definitions 4.2 and 4.4 are equivalent. Indeed, if $f \in C(\mathbb{Z}_p, \mathbb{Z}_p)$ is locally representable by a restricted power series in $\mathbb{Z}_p[[x]]$, then it is analytic in the sense of Definition 4.2. Conversely, let as assume that $f \in C(\mathbb{Z}_p, \mathbb{Z}_p)$ is locally represented by a restricted power series $F \in \mathbb{Q}_p[[x]]$ of the form (4.1), whose coefficients satisfy (4.2), on a disc of radius $1/p^N$, $N \geq 1$. Thus we have that $f(x) = F(x - k)$ for some $k \in \{0, \dots, p^N - 1\}$ for all $x \in k + p^N \mathbb{Z}_p$. Without loss of generality, let us assume that $k = 0$. Computing the value of f at $x = 0$, we obtain that $f(0) = a_0$, and so $a_0 \in \mathbb{Z}_p$, it being in the range of f. By the non-Archimedean property of the norm, the function $f(x) - a_0$ ranges in $\mathbb{Z}_p$ also, and we have that $f(x) - a_0 = x g(x)$ for $g \in C(\mathbb{Z}_p, \mathbb{Z}_p)$, and this g is represented by the series $\sum_{j=1}^{\infty} a_j x^{j-1}$ on the said disc. Proceeding by induction, we then show that all the a_js are in $\mathbb{Z}_p$.

   With this in mind, we can say that a function $f : \mathbb{Z}_p \to \mathbb{Z}_p$ is analytic in the sense of [36] (cf. with [42], p. LG 2.4) if, and only if, it is analytic of level m for some m. $\qquad\square$

Example 4.6 Given a subspace $X \subset \mathbb{Q}_p$, a function $f : X \mapsto \mathbb{Q}_p$ is said to be locally constant on X if for every $\alpha \in X$, there exists n such that f is constant on the disc $D_X(\alpha, p^{-n}) = (\alpha + p^n \mathbb{Z}_p) \cap X$. Any locally constant function $f : \mathbb{Z}_p \mapsto \mathbb{Z}_p$ is analytic. A typical example of that is the characteristic function $\chi_{D(0,1/p)}$ of the disc in $\mathbb{Z}_p$ with center at zero and radius $1/p$, an analytic function of level 1.

Mahler's expansion of this characteristic function (see Theorem 3.2) is given by

$$\chi_{D(0,1/p)}(x) = \sum_{n=0}^{\infty} a_n \begin{pmatrix} x \\ n \end{pmatrix},$$

where

$$a_n = (-1)^n \sum_{j=0}^{[\frac{n}{p}]} (-1)^{jp} \begin{pmatrix} n \\ jp \end{pmatrix},$$

and so, for instance,

$$\chi_{D(0,1/2)}(x) = 1 + \sum_{n=1}^{\infty} (-1)^n 2^{n-1} \begin{pmatrix} x \\ n \end{pmatrix} \text{ in } \mathbb{Z}_2,$$

$$\chi_{D(0,1/3)}(x) = 1 + \sum_{n=1}^{\infty} (-1)^n 2 \cos\left(\frac{n\pi}{6}\right) 3^{-1+\frac{n}{2}} \begin{pmatrix} x \\ n \end{pmatrix} \text{ in } \mathbb{Z}_3.$$

In general, the p-adic coefficient a_n can be expressed in terms of a primitive p-root of unity ζ_p by the formula

$$a_n = p^{-1} \sum_{k=1}^{p} (\zeta_p^k - 1)^n.$$

Notice that this expression shows the rational integer a_n as a sum of elements in $\mathbb{Q}_p(\zeta_p)$. In Example 3.10, we proved that $\zeta_p^k - 1$ has p-adic norm $1/p^{p-1}$. Therefore,

$$\|a_n\|_p \leq p^{-n(p-1)+1},$$

and we conclude that the Mahler power series of $\chi_{D(0,1/p)}$ is a restricted power series in $\mathbb{Z}_p$. Further, we see that $\{\|a_n\|_p\}$ converges to zero faster than $p^{-n/p}$, and this growth can be used to conclude that $\chi_{D(0,1/p)}(x)$ is an analytic function [36, p. 85]. But observe that drawing that conclusion on the basis of properties of the a_ns is not an elementary result. $\qquad\square$

The example above illustrates the difficulties we face when dealing with issues of analyticity. The polynomial functions $\begin{pmatrix} x \\ n \end{pmatrix}$ are very natural candidates to use when we wish to approximate uniformly a continuous function f in an optimal manner. But if f is analytic, that fact is not easily captured in properties of the coefficients of the approximation.

Example 4.7   The projection

$$\mathbb{Z}_p \xrightarrow{\pi_n} \mathbb{Z}_p$$
$$\alpha \mapsto c_n$$

onto the n-th coefficient c_n in the p-adic expansion (2.4) of the p-adic integer α is a locally constant function on $\mathbb{Z}_p$. Notice that π_n is a linear combination of the locally constant characteristic functions of the discs $a + p^{n+1}\mathbb{Z}_p$, $0 \le a \le p^{n+1} - 1$. $\Box$

Example 4.8   For any $\alpha \in \mathbb{Q}_p$, there exists a unique representation

$$\alpha = \sum_{j \ge n} \omega_j(\alpha)p^j \,,$$

where all the coefficients $\omega_j(\alpha)$ are $(p-1)$-roots of unity. This is the Teichmüller representation of α, and in some sense, it is a more natural choice than that given by the representation (2.4). See its essential use in the proof of Theorem 9.2 below.

In order to define the functions $\alpha \mapsto \omega_j(\alpha)$ in the Teichmüller representation of α, it suffices to do so for α a p-adic integer. So we proceed to define the functions $\omega_j : \mathbb{Z}_p \mapsto \mathbb{Z}_p$ for $j = 0, 1, \ldots$.

Given $\alpha \in \mathbb{Z}_p$, we have $\alpha = \sum_{j=0}^{\infty} c_j p^j$ where the c_js are all rational integers in the range $0 \le c_j < p$. Let us consider the root α_{c_0} of the polynomial $x^p - x$ associated to the first coefficient c_0, that is to say, $\alpha_{c_0} = \lim c_0^{p^n}$. By our discussion in §2.5, we have either $\alpha_{c_0} = 0$ when $c_0 = 0$ or α_{c_0} a $(p-1)$-root of unity in $\mathbb{Z}_p$ when $c_0 \ne 0$. We define the Teichmüller character function ω by

$$\mathbb{Z}_p \xrightarrow{\omega} \mathbb{Z}_p$$
$$\alpha \mapsto \alpha_{c_0} \,.$$

It has the properties that

$$\omega(\alpha\beta) = \omega(\alpha)\omega(\beta)$$

and that

$$\|\omega(\alpha + \beta) - (\omega(\alpha) + \omega(\beta))\|_p < 1\,,$$

respectively.

The zeroth Teichmüller coefficient function is defined by

$$\omega_0(\alpha) = \omega(\alpha)\,.$$

We have the estimate

$$\|\alpha - \omega_0(\alpha)\|_p \leq \max\{\|\alpha - c_0\|_p\,, \|c_0 - \omega_0(\alpha)\|_p\} < 1\,,$$

and so $\alpha \equiv \omega_0(\alpha)$ mod p. The remaining Teichmüller coefficient functions can then be defined by the recursion formula

$$\omega_n(\alpha) = \omega\left(\frac{\alpha - \sum_{j=0}^{n-1} \omega_j(\alpha)p^j}{p^n}\right)\,, \quad n = 1, 2, \ldots\,.$$

We then have that $\alpha \equiv \sum_{j=0}^{n} \omega_j(\alpha)p^j$ mod p^{n+1}.

As is the case of the projection functions π_n in Example 4.7, all the Teichmüller coefficient functions $\alpha \mapsto \omega_n(\alpha)$ are locally constant. $\qquad\Box$

4.1 Strassmann's theorem

Let us consider a nontrivial restricted power series

$$F(x) = \sum_{j=0}^{\infty} a_j x^j \in \mathbb{Q}_p[[x]]\,, \quad \lim_{j \to \infty} \|a_j\|_p = 0\,, \tag{4.3}$$

as a function with domain of definition given by $\mathbb{Z}_p$. We have that $\|a_j\|_p \leq M$ for some constant M. In fact, the optimal constant M is equal to $\max_j\{\|a_j\|_p\}$, which is achieved as the p-adic norm of some of the coefficients. Let N be an integer such that

$$\|a_N\|_p = \max\{\|a_j\|_p\}\,, \quad \|a_j\|_p < \|a_N\|_p \text{ for all } j > N. \tag{4.4}$$

We shall need the following theorem of Strassmann [47] later on. We present its proof mostly for completeness, but also as a way to illustrate the contrast between p-adic and the usual real analysis.

Theorem 4.9 *Consider a nontrivial restricted power series of the form (4.3), and view it as a function $F : \mathbb{Z}_p \to \mathbb{Q}_p$. Let N be an integer such that (4.4) holds. Then F has at most N roots in $\mathbb{Z}_p$.*

Proof. The result uses the completeness of $\mathbb{Z}_p$, and the non-Archimedean property of the p-adic norm $\| \ \|_p$. It follows via an induction argument on N.

Indeed, if $N = 0$ and $F(x_0) = 0$ then

$$a_0 = -\sum_{j=1}^{\infty} a_j x_0^j \, ,$$

and by the non-Archimedean property of $\| \ \|_p$, we obtain the contradictory inequality

$$\|a_0\|_p = \left\| \sum_{j=1}^{\infty} a_j x_0^j \right\|_p \leq \max_{j \geq 1} \left\| a_j x_0^j \right\|_p \leq \max_{j \geq 1} \|a_j\|_p < \|a_0\|_p \, .$$

Suppose now that $N > 0$, and let x_0 be a root of $F(x)$. Then

$$F(x) = F(x) - F(x_0) = \sum_{j \geq 1} a_j (x^j - x_0^j) = (x - x_0) \sum_{j \geq 1} \sum_{k=0}^{j-1} a_j x^k x_0^{j-1-k} \, .$$

We can interchange the order of summation to obtain that

$$G(x) := \sum_{j \geq 1} \sum_{k=0}^{j-1} a_j x^k x_0^{j-1-k} = \sum_{k=0}^{\infty} \left(\sum_{l \geq 0} a_{k+1+l} x_0^l \right) x^k = \sum_{k \geq 0} g_k x^k \, ,$$

where we let the last equality define the coefficient g_k. Thus, we see that $G(x)$ is also a restricted power series whose coefficients are all p-adically bounded by $\|a_N\|_p$, we have that $\|g_{N-1}\|_p = \|a_N\|_p$ and $\|g_j\|_p < \|g_{N-1}\|_p$ for all $j \geq N$. By induction, the function $G(x)$ has at most $N - 1$ roots in $\mathbb{Z}_p$, and so

$$F(x) = (x - x_0)G(x)$$

can have at most N roots altogether. $\qquad \square$

Our use of this result will be in the form of an easy consequence: if two power series of the indicated form agree over infinitely many p-adic integers, then they must be identical.

5

Arithmetic differential operators on $\mathbb{Z}_p$

In going further, we now make a fundamental use of Theorem 2.12 to introduce the notion of arithmetic differential operators over the ring $\mathbb{Z}_p$. This is the essential idea in our work here. In the following chapter, we shall present the theory of these operators in general, discussing the jet spaces whose global sections give rise to them. We shall even expand our explanations to outline the general theory in the case of multiple primes, something that we start also in a simplified manner here. But when we revisit the study of these operators in the remaining chapters, there will be little that we will do beyond our discussion of them over the ring $\mathbb{Z}_p$ until the very end. In the very last chapter, we compare the behaviour of these operators over the p-adic integers with their behaviour over its unramified completion.

The idea leading to arithmetic operators embodies a radically different philosophy from that used up until now. This philosophy arises naturally when looking at the p-adic numbers in the setting of the analytic functions of the previous section. For we treat p-adic numbers now as functions, albeit functions on a space of "dimension zero." These functions all admit the representation (2.4), and so they ought to be considered analytic since these representations are convergent power series in p. Once we agree with the idea of the p-adic numbers as "analytic" functions, it is then only natural to define their derivatives, the *arithmetic derivatives* of our work. In this sense, the theory departs from the idea that the rings we use —in most of our analysis the ring $\mathbb{Z}_p$— are to be considered as ring of functions. This point of view leads to a very strong tie between algebraic geometry and algebraic number theory.

The idea of treating numbers as functions is classical, and goes back to at least R. Dedekin, W. Weber, and D. Hilbert. But the arithmetic differentiation that we discuss here is due to A. Buium in his work exploiting Fermat quotients [8] (see also [10, 6]). This theory has been extended recently, and we now have a notion of arithmetic partial differential equations available [15, 16, 17], the implications of which are currently under investigation.

Let $c_p(x, y)$ be the element of the polynomial ring $\mathbb{Z}[x, y]$ given by

$$c_p(x, y) = \frac{x^p + y^p - (x + y)^p}{p}. \tag{5.1}$$

Let A be a ring and B be an algebra over A. If $x \in A$, we denote by x its image in B also. We say that a map $\delta_p : A \to B$ is a p-derivation if $\delta_p(1) = 0$, and

$$\begin{aligned} \delta_p(x + y) &= \delta_p(x) + \delta_p(y) + c_p(x, y), \\ \delta_p(xy) &= x^p \delta_p(y) + y^p \delta_p(x) + p \delta_p(x) \delta_p(y), \end{aligned} \tag{5.2}$$

for all $x, y \in A$, respectively.

The first of the identities above indicates that the p-derivation δ_p is additive modulo the zeroth order term $c_p(x, y) \in \mathbb{Z}[x, y]$. The second of the identities is a sort of nonlinear Leibniz's rule mod p, where the function factors in the two summands in the right side are composed with the p-th power Frobenius map.

Any p-derivation δ_p defines a ring homomorphism $\phi_p : A \to B$ by the expression

$$\phi_p(x) = x^p + p \delta_p(x). \tag{5.3}$$

For by (5.2), we have that

$$\begin{aligned} \phi_p(x + y) &= (x + y)^p + p \delta_p(x) + p \delta_p(y) + p c_p(x, y) \\ &= \phi_p(x) + \phi_p(y), \end{aligned}$$

while

$$\begin{aligned} \phi_p(xy) &= x^p y^p + p x^p \delta_p(y) + p y^p \delta_p(x) + p^2 \delta_p(x) \delta_p(y) \\ &= (x^p + p \delta_p(x))(y^p + p \delta_p(y)) = \phi_p(x) \phi_p(y). \end{aligned}$$

The homomorphism ϕ_p commutes with the p-derivation δ_p.

Conversely, given a homomorphism $\phi_p : A \mapsto B$, the expression

$$\delta_p(x) = \frac{\phi_p(x) - x^p}{p} \tag{5.4}$$

defines a δ_p-derivation, and we have that

$$\phi_p \delta_p(x) = \phi_p \left(\frac{\phi_p(x) - x^p}{p} \right) = \frac{\phi_p(\phi_p(x)) - (\phi_p(x))^p}{p} = \delta_p \phi_p(x).$$

We say that the p-derivation δ_p and the homomorphism ϕ_p are *associated* to each other.

Definition 5.1   A δ_p-*ring* A is a ring A equipped with a p-derivation $\delta_p : A \to A$. A *morphism of* δ_p-*rings* is a ring homomorphism that commutes with the δ_p-derivations in each of the rings. A δ_p-ring B is said to be a δ_p-*ring over the* δ_p-*ring* A if it comes equipped with a δ_p-ring homomorphism $A \to B$. A δ_p-ring A is a δ_p-*subring* of the δ_p-ring B if A is a subring of B such that $\delta_p A \subset A$. $\square$

We now study this general situation over the specific case of the ring $\mathbb{Z}_p$. In this ring, the unique homomorphism ϕ_p that lifts the p-th power Frobenius isomorphism is the identity. Thus, a p-derivation on $\mathbb{Z}_p$ is associated to $\phi_p(x) = x$.

In fact, this assertion can be obtained independently. By Theorem 2.12, $x^p \equiv x \bmod p$, and so we can consider the Fermat quotient

$$\delta_p(x) = \frac{x - x^p}{p} . \tag{5.5}$$

to define a p-derivation δ_p on $\mathbb{Z}_p$.

Theorem 5.2   *The ring* $\mathbb{Z}_p$ *carries a unique* p-*derivation* δ_p *that is defined by the expression* (5.5).

Proof. By the observation made above, the set of p-derivations on $\mathbb{Z}_p$ are in 1-to-1 correspondence with ring automorphisms of $\mathbb{Z}_p$. Since the identity is the only such homomorphism, the assertion follows. $\square$

Since p is fixed,when referring to the p-derivation (5.5) we shall often write δx instead of $\delta_p(x)$, except when confusion could arise.

We denote by δ^i the i-th iterate of δ. We use the convention that $\delta^0 a = a$.

Definition 5.3   A function $f : \mathbb{Z}_p \to \mathbb{Q}_p$ is called an *arithmetic differential operator of order* m, or a δ_p-*function* of order m, if there exists a restricted power series $F \in \mathbb{Q}_p[[x_0, x_1, \ldots, x_m]]$ such that

$$f(a) = F(a, \delta a, \ldots, \delta^m a) \tag{5.6}$$

for all $a \in \mathbb{Z}_p$. $\square$

We shall refer to an *arithmetic differential operator* of order m simply as a δ-differential operator of order m also, or as δ-*function* of order m.

Lemma 5.4 *If* $f, g : \mathbb{Z}_p \to \mathbb{Q}_p$ *are arithmetic differential operators of orders* m *and* n, *respectively, then* $f \circ g$ *is an arithmetic differential operator of order* $m + n$. *In particular, the set of arithmetic differential operators* $f : \mathbb{Z}_p \to \mathbb{Z}_p$ *is closed under compositions.*

Proof. This is elementary. $\square$

Example 5.5   Using a slightly more general context, the theory of arithmetic differential operators in [6, 8] gives rise to several interesting number theoretic locally constant functions that have nice representations as arithmetic differential operators of low order. These have been the main source of motivation for our work, and so we describe one of them briefly here, somewhat out of context. As we have indicated already, the general theory will be outlined in the following chapter. Lastly, at the very end of our work, we present other examples that sparked our interest.

Again, since functions of the type (5.6) are continuous and $\mathbb{Z}_p$ is compact, via a scaling we can reduce our consideration to the case where F is a restricted power series with coefficients in $\mathbb{Z}_p$. We can extend Definition 5.3 by considering functions $f : \mathbb{Z}_p^N \to \mathbb{Z}_p$ that can be represented as in (5.6) with an $F \in \mathbb{Z}_p[[x_0, \ldots, x_m]]$, but where now each x_j is an N-tuple of variables. We call these *arithmetic differential operators of order* m also. If X is an affine scheme embedded* into the affine N-space over $\mathbb{Z}_p$, we let $X(\mathbb{Z}_p) \subset \mathbb{Z}_p^N$ be the natural inclusion at the level of $\mathbb{Z}_p$-points. We then call a function $X(\mathbb{Z}_p) \to \mathbb{Z}_p$ an *arithmetic differential of order* m if it can be extended to an arithmetic differential operator $\mathbb{Z}_p^N \to \mathbb{Z}_p$ of order m. This suffices for now to introduce the Legendre symbol function as discussed in [10].

Let X be the multiplicative group scheme over $\mathbb{Z}_p$ embedded into the affine plane $\operatorname{Spec} \mathbb{Z}_p[v, w]$ via the map $u \mapsto (u, u^{-1})$. Then we have that $X(\mathbb{Z}_p) = \mathbb{Z}_p^\times$, and we can talk about arithmetic differential operators $\mathbb{Z}_p^\times \to \mathbb{Z}_p$ of order m. The Legendre symbol turns out to be a rather remarkable example of one such operator, as we see here.

If p is an odd prime, the Legendre symbol

$$f : \mathbb{Z}_p^\times \to \mathbb{Z}_p$$

---

* For the convenience of the reader not familiar with this or several other algebraic concepts that we use, we recall them in §6.1 below.

is the function $a \mapsto (a/p)$, defined by

$$\left(\frac{a}{p}\right) = \begin{cases} 1 & \text{if } a \text{ is a quadratic residue mod } p, \\ -1 & \text{if } a \text{ is a quadratic nonresidue mod } p. \end{cases}$$

for all a such that $(a, p) = 1$. This function is the δ-differential operator of order 1 on $\mathbb{Z}_p^{\times}$ given by

$$
\begin{aligned}
\left(\frac{a}{p}\right) &= a^{\frac{p-1}{2}} \left(1 + \sum_{n=1}^{\infty} (-1)^{n-1} \frac{(2n-2)! p^n}{2^{2n-1}(n-1)! n!} (\delta a)^n a^{-pn} \right) \\
&= a^{\frac{p-1}{2}} \left(1 + p \frac{\delta a}{a^p} \right)^{\frac{1}{2}}.
\end{aligned}
\tag{5.7}
$$

(See Theorem 3.1.) Indeed, the three expressions involved in this equality are all congruent to $a^{\frac{p-1}{2}}$ mod p, and by (5.5), the square of the right side is just

$$a^{p-1}\left(1 + p\frac{\delta a}{a^p}\right) = a^{p-1}\frac{a^p + p\delta a}{a^p} = 1.$$

We see then that if we define the series

$$F(x, y, \xi, \zeta) = x^{\frac{p-1}{2}} \left(1 + \sum_{n=1}^{\infty} (-1)^{n-1} \frac{(2n-2)! p^n}{2^{2n-1}(n-1)! n!} \xi^n y^{pn} \right)$$

as a restricted power series of four variables, independent of ζ, we have that

$$F(a, a^{-1}, \delta a, \delta a^{-1}) = a^{\frac{p-1}{2}} \left(1 + p\frac{\delta a}{a^p} \right)^{\frac{1}{2}}.$$

Notice that the Legendre symbol is a locally constant function of level 1, that is to say, constant on discs of radius $1/p$ in its domain of definition. It is therefore analytic. We shall see below that this latter fact holds in general for any arithmetic differential operator on $\mathbb{Z}_p$. $\quad \square$

The theory of arithmetic differential equations we just introduced is intrinsically nonlinear. For let us recall that in classical analysis, a linear differential operator of order zero on a manifold M is just given by a function on M, which is thought of as a multiplication operator on the space of functions itself. Then a linear differential operator P of order n is defined to be a linear operator on the space of functions such that the commutator $[P, f] = Pf - fP$ is a differential operator of order $n-1$ for any zeroth order operator f. The application of this line of reasoning to arithmetic differential operators is fruitless, and shows immediately their intrinsic nonlinear nature. The proper analogue of linearity is otherwise.

Indeed, if $f : \mathbb{Z}_p \to \mathbb{Z}_p$ is a $\mathbb{Z}_p$-valued function on $\mathbb{Z}_p$ that we think of as the multiplication operator given by $a \mapsto f(a)a$, using the multiplicative property of (5.5) encoded in the second identity in (5.2), we have that

$$[\delta, f](a) = \delta(f(a)a) - f(a)\delta a = a^p \delta(f(a)).$$

Thus, the commutator $[\delta, f]$ is the product of $\delta(f)$ and the p-th power mapping. The alter egos of the n-th order linear differential operators in classical analysis would be compositions of iterates of the operator δ and the p-th power Frobenius map up to order n, which makes of them highly nonlinear expressions in the argument anyway: no commutator $[F, f]$ of a nontrivial δ-operator F and a nontrivial function f can be linear in the argument.

In appropriate cases, there is a convenient notion of *linearity* in the context of arithmetic differential operators, a notion that was already present in [6, 8]. It refers to a group homomorphism property in cases where the domain of the arithmetic differential operator makes it possible to talk about such. We outline this notion here. Details will be given in §6.3.

We place our discussion in the context introduced in Example 5.5 above. We recall that if X is an affine scheme embedded into the affine N-space over $\mathbb{Z}_p$, a function $X(\mathbb{Z}_p) \to \mathbb{Z}_p$ is an arithmetic differential if it can be extended to an arithmetic differential operator $\mathbb{Z}_p^N \to \mathbb{Z}_p$. Suppose that the X in this general set-up were to be a commutative group scheme G over $\mathbb{Z}_p$. Then we could define a "linear arithmetic differential operator" on G to be a differential operator

$$G(\mathbb{Z}_p) \to \mathbb{Z}_p$$

on G that is also a group homomorphism. Here, the range $\mathbb{Z}_p$ in the right side is viewed with its additive group structure. By making this "set-theoretical" definition one that is "scheme-theoretical" varying the ring $\mathbb{Z}_p$, we would obtain the notion of *linear arithmetic differential operator* of [6, 8] that was alluded to above. With this definition in place, we recover the familiar property that "linear operators" have, namely that the difference of solutions to homogeneous equations is again a solution.

Example 5.5 illustrates on $\mathbb{Z}_p^\times$ the type of questions we would like to ask of δ-functions over $\mathbb{Z}_p$ with as much generality as possible. The Legendre symbol admits a representation as an arithmetic differential operator of order 1. We shall prove that an analytic function on $\mathbb{Z}_p$ of level m can be represented as an arithmetic differential operator whose order is m.

Let us reiterate for the moment that when thinking of p-adic numbers as functions, these functions are in fact analytic. This comes about via the natural analogy that the representation (2.4) generates. It is an irony that this analogy is at best somewhat incomplete: though we know of analytic functions [14], we do not yet have the proper notion of what a *continuous arithmetic function* ought to be.

Remark 5.6 Buium's general theory of arithmetic differential operators [6, 8] is developed over the ring $R := R_p := \widehat{\mathbb{Z}}_p^{ur}$, the completion of the maximal unramified extension of $\mathbb{Z}_p$, in the role of our $\mathbb{Z}_p$ here. Let $k = R/pR$ be the residue field, and let $\mu(R)$ be the multiplicative group of roots of unity in R. We recall that the reduction mod p mapping

$$\mu(R) \to k^\times$$

defines an isomorphism whose inverse is the Teichmüller lift. Any element of the ring R can be represented uniquely as a series $\sum_{i=0}^\infty \zeta_i p^i$, where $\zeta_i \in \mu(R) \cup \{0\}$. Using this representation of the elements of R, we shall see that there exists a unique ring isomorphism

$$\phi : R \to R$$

that lifts the p-th power Frobenius isomorphism on the residue field k, and for $\zeta \in \mu(R)$, we have that $\phi(\zeta) = \zeta^p$. This is the associated homomorphism (5.3) of the p-derivation on the ring R giving rise to the general theory (see Theorem 6.9 below). $\square$

5.1 Multiple primes I

The theory of arithmetic differential operators introduced above has been extended to the case where there is more than one prime [17]. Elaborating on this idea requires the use of more complicated rings than those that play a role in our work on the single prime case. We outline some of the issues related to this next.

Let us begin by revisiting the presentation above and reformulating it slightly, reversing somewhat the order followed when introducing the concept of arithmetic derivative δ_p in (5.5). We now begin our work at the level of the ring $\mathbb{Z}$ instead of $\mathbb{Z}_p$, and replace the use of Theorem 2.12 on $\mathbb{Z}_p$ directly by that of Fermat's little theorem at the level of $\mathbb{Z}$. We then use the fact that the identity is the unique automorphism ϕ in the latter ring, and so given a prime p, Fermat's little theorem ensures the existence of a unique δ_p derivation on $\mathbb{Z}$ associated to this homomorphism $\phi_p = \phi$. Now in §2.2 the ring $\mathbb{Z}_p$ is presented as the p-adic completion of $\mathbb{Z}$ at the ideal $(p) \subset \mathbb{Z}$ (cf. definition in §6.1). The homomorphism ϕ_p uniquely extends by continuity to $\mathbb{Z}_p$, and the condition $\phi_p(a) \equiv a^p \bmod p$ in $\mathbb{Z}_p$ will hold for the extension by continuity also. We thus obtain a unique p-derivation δ_p on $\mathbb{Z}_p$ associated to ϕ_p. Notice that under this presentation, the analogy between $(\mathbb{C}[x], d/dx)$ and $(\mathbb{Z}, \delta)$ discussed in the introduction becomes clear: the fundamental theorems of algebra and arithmetic make the linear polynomials in $\mathbb{C}[x]$ correspond to the primes in $\mathbb{Z}$, and when $p > 2$, the set $\{-1, 0, 1\} \subset \mathbb{Z}$ is the analogue of the field of constant polynomials in $\mathbb{C}$, for these integers are the only solutions of the equation $\delta_p n = 0$ in $\mathbb{Z}$. When passing to $\mathbb{Z}_p$ by continuity, the set of constants is enlarged to encompass the roots of unity of order prime to p.

A theory involving at least a second prime forces us to consider as the simplest ring to use one that carries sufficiently many automorphisms. The ring $\mathbb{Z}$ is unsatisfactory in this respect, and the procedure above requires modifications, which we present now. In this way, we point towards the introduction of arithmetic operators in the case of multiple primes, and the general difficulties we face when doing so. The basic ring to use is essentially obtained from $\mathbb{Z}$ by adjoining at least one root of unity, and then taking its completion relative to the prime ideal generated by the set of primes we are using. The details illustrate the difficulty that the reader will no doubt see clearly: how do we make two or more primes interact with each other in a reasonable way.

Given two distinct primes p and q in $\mathbb{Z}$, we consider the polynomial $c_{p,q}$ in the ring $\mathbb{Z}[x_0, x_1, x_2]$ defined by

$$c_{p,q}(x_0, x_1, x_2) = \frac{c_q(x_0^p, px_1)}{p} - \frac{c_p(x_0^q, qx_2)}{q} - \frac{\delta_p q}{q} x_2^p + \frac{\delta_q p}{p} x_1^q, \quad (5.8)$$

where for each prime p, $c_p \in \mathbb{Z}[x, y]$ and δ_p are the polynomial and p-derivation in (5.1) and (5.5), respectively. Notice that this polynomial $c_{p,q}$ lies in the ideal $(x_0, x_1, x_2)^{\min\{p,q\}} \subset \mathbb{Z}[x_0, x_1, x_2]$.

We now have the following [17].

Definition 5.7 Let $\mathcal{P} = \{p_1, \ldots, p_d\}$ be a finite set of primes in $\mathbb{Z}$. A $\delta_{\mathcal{P}}$-*ring* is a ring A endowed with p_k-derivations $\delta_{p_k} : A \to A$, $k = 1, \ldots, d$, such that

$$\delta_{p_k} \delta_{p_l} a - \delta_{p_l} \delta_{p_k} a = c_{p_k, p_l}(a, \delta_{p_k} a, \delta_{p_l} a) \quad (5.9)$$

for all $a \in A$, $k, l = 1, \ldots, d$. A *homomorphism of $\delta_{\mathcal{P}}$-rings* A and B is a homomorphism of rings $\varphi : A \to B$ that commutes with the p_k-derivations in A and B, respectively. $\qquad\square$

Let $\mathcal{P} = \{p_1, \ldots, p_d\}$ be a finite set of primes in $\mathbb{Z}$, and A be a $\delta_{\mathcal{P}}$-ring with p_k-derivations δ_{p_k}, as in the definition. For each index k in the range $1, \ldots, d$, we let ϕ_{p_k} denote the homomorphism (5.3) associated with the p_k-derivation δ_{p_k}. Then condition (5.9) implies the commutativity relations $\phi_{p_k} \phi_{p_l} = \phi_{p_l} \phi_{p_k}$. Indeed, for all $a \in A$, we obtain that

$$
\begin{aligned}
\phi_{p_k} \phi_{p_l}(a) &= \phi_{p_k}\left(a^{p_l} + p_l \delta_{p_l} a\right) \\
&= a^{p_l p_k} + p_k \delta_{p_k} a^{p_l} + p_l (\delta_{p_l} a)^{p_k} + + p_k p_l \delta_{p_k} \delta_{p_l} a \\
&= \phi_{p_l} \phi_{p_k}(a) + p_k \delta_{p_k} a^{p_l} - p_l \delta_{p_l} a^{p_k} + p_l (\delta_{p_l} a)^{p_k} \\
&\quad - p_k (\delta_{p_k} a)^{p_l} + p_k p_l (\delta_{p_k} \delta_{p_l} a - \delta_{p_l} \delta_{p_k} a) \\
&= \phi_{p_l} \phi_{p_k}(a),
\end{aligned}
$$

where the last equality follows by substituting $c_{p_k, p_l}(a, \delta_{p_k} a, \delta_{p_l} a)$ for $\delta_{p_k} \delta_{p_l} a - \delta_{p_l} \delta_{p_k} a$, and simple cancellations. In turn, the commutation of the associated homomorphisms implies the commutation relations $\phi_{p_k} \delta_{p_l} a = \delta_{p_l} \phi_{p_k} a$.

Conversely, suppose the homomorphisms ϕ_k, ϕ_l satisfy the commutation relations above, and that the p_ks are nonzero divisors in A. Then the conditions (5.9) hold, and we have that

$$
\phi_{p_k} \delta_{p_l} a = \delta_{p_l} \phi_{p_k} a
$$

for all $a \in A$.

A nontrivial example of a $\delta_{\mathcal{P}}$-ring ring requires it to carry sufficiently many automorphisms that can be the homomorphisms associated with the various arithmetic derivations being considered. Out of necessity, the basic example of ring to use must be larger than $\mathbb{Z}$ and with sufficiently many automorphisms.

Example 5.8 Let $S \subset \mathbb{Z}$ be a multiplicative system of integers coprime to $p_1, \ldots, p_d$, and let $\mathbb{Z}_S = S^{-1}\mathbb{Z}$ be the corresponding ring of fractions. For a given integer m that is coprime to $p_1, \ldots, p_d$, we consider the m-th root of unity $\zeta_m = \exp\left(\frac{2\pi\sqrt{-1}}{m}\right)$, and the polynomial ring $A = \mathbb{Z}_S[\zeta_m]$ in ζ_m. Notice that ζ_m is a root of the separable polynomial $x^m - 1$, which has m distinct roots in the algebraic closure of $\mathbb{Q}$, all of them roots of unity. We have that A is contained in the field $\mathbb{Q}(\zeta_m)$, and this extension of $\mathbb{Q}$ is Galois and commutative.

If ζ is a primitive m-root of unity and σ is an automorphism of $\mathbb{Q}(\zeta_m)$, then $\sigma(\zeta)$ is another m-root of unity, which therefore must be of the form $\zeta^{n(\sigma)}$ for some integer $n = n(\sigma)$. The map

$$G(\mathbb{Q}(\zeta_m)/\mathbb{Q}) \ni \sigma \mapsto n(\sigma) \mod m$$

defines an isomorphism between the Galois group $G(\mathbb{Q}(\zeta_m)/\mathbb{Q})$ and $(\mathbb{Z}/m\mathbb{Z})^\times$, respectively.

The primes $p_1, \ldots, p_d$ under consideration define classes in $(\mathbb{Z}/m\mathbb{Z})^\times$. We let $\phi_{p_1}, \ldots, \phi_{p_d}$ be the corresponding elements of the Galois group $G(\mathbb{Q}(\zeta_m)/\mathbb{Q})$. Then $\phi_{p_k}(a) \equiv a^{p_k} \mod p_k$ for $k = 1, \ldots, d$ and $a \in A$, and A becomes a $\delta_{\mathcal{P}}$-ring with respect to the p_k-derivations δ_{p_k} associated to the homomorphisms ϕ_{p_k}, $k = 1, \ldots, d$.

All the p_k-adic completions (see §6.1 for definition) $A^{\widehat{(p_k)}}$ at the ideal $(p_k) \subset \mathbb{Z}$ are $\delta_{\mathcal{P}}$-rings in a natural way. For ϕ_{p_l} extends to $A^{\widehat{(p_k)}}$ by continuity, the condition $\phi_{p_k} a \equiv a^{p_k} \mod p_k$ in $A^{\widehat{(p_k)}}$ holds by continuity, and the condition $\phi_{p_l} a \equiv a^{p_l} \mod p_l$ in $A^{\widehat{(p_k)}}$ holds because p_l is invertible in $A^{\widehat{(p_k)}}$. $\qquad\square$

Example 5.9 Let us take a closer look at the ring $\mathbb{Z}_S$ of the ring $\mathbb{Z}_S[\zeta_m]$ in the example above. We can take $S = \cap_{j=1}^d(\mathbb{Z} \setminus (p_j))$ and the ring of fractions $\mathbb{Z}_S$ will then be

$$\mathbb{Z}_{\mathcal{P}} = \{m/n : m, n \in \mathbb{Z} \text{ such that } (n, p_j) = 1 \text{ for } j = 1, \ldots, d\}.$$

By Fermat's little theorem, if $a \in \mathbb{Z}_{\mathcal{P}}$, then we have that $a \equiv a^{p_j} \mod p_j$, $j = 1, \ldots, d$. So we obtain a $\delta_{\mathcal{P}}$-ring structure on $\mathbb{Z}_{\mathcal{P}}$. This ring is in fact the intersection $\cap_{j=1}^d \mathbb{Z}_{(p_j)}$ of the localizations (see §6.1 for definition; cf. with Example 6.1) $\mathbb{Z}_{(p_j)}$ of $\mathbb{Z}$ at the ideals generated by the various primes under consideration.

If we were to develop a multiple primes theory parallel to the one we study here, $\mathbb{Z}_{\mathcal{P}}$ would have to take on the role of the ring $\mathbb{Z}$. Let us consider the ideal $i_{\mathcal{P}} = (p_1, \ldots, p_d)$ generated by the primes in $\mathcal{P}$. This ideal induces a topology on $\mathbb{Z}_{\mathcal{P}}$, and the completion $\widehat{\mathbb{Z}}_{\mathcal{P}}^{i_{\mathcal{P}}}$ of $\mathbb{Z}_{\mathcal{P}}$ at $i_{\mathcal{P}}$ is the ring that then takes on the role of $\mathbb{Z}_p$ (again, see §6.1 for the definitions of the unexplained terms used here). This general approach is a natural idea to adopt [17]. $\qquad\square$

6

A general view of arithmetic differential operators

We present now a fuller view of arithmetic differential operators by out-lining their general theory. We shall not use this theory with this degree of generality anywhere else in this monograph, and the additional details that we provide later on are only there to make of our treatment a self contained one when emphasizing the different flavor these operators have when the ring $\mathbb{Z}_p$ is enlarged.

Arithmetic differential operators over $\mathbb{Z}_p$ have a special behaviour in comparison to their behaviour when the ground ring is larger, as in the general theory. In the general theory [6, 8], the role of the the ring $\mathbb{Z}_p$ is played by by the completion $\widehat{\mathbb{Z}}_p^{ur}$ of the maximum unramified extension of $\mathbb{Z}_p$, and the p-derivation δ is now given by

$$
\begin{array}{ccc}
\widehat{\mathbb{Z}}_p^{ur} & \xrightarrow{\delta} & \widehat{\mathbb{Z}}_p^{ur} \\
a & \mapsto & \dfrac{\phi(a) - a^p}{p} \,,
\end{array}
\tag{6.1}
$$

where ϕ is the unique lift of Frobenius to $\widehat{\mathbb{Z}}_p^{ur}$. The differences that result between the ensuing two theories over these rings are analogous to the differences between number theoretic statements about finite fields, and algebraic geometric statements over their algebraic closures. The theory as we discuss it pertains a single prime p, as in [6, 5]. The extended theory for several primes may be found in [17]. We shall provide some details of it also, in §6.4, outlining further what was started in §5.1. The reader interested in a quick review of the entire theory may find it convenient to consult the survey article [13].

As a guiding principle, let us observe that arithmetic differential operators were originally defined paralleling the theory of differential operators on bundles over a manifold. In order to clarify this analogy, we recall the latter briefly. The clasic definition of differential operators

65

is based on the theory of jet bundles [40], all of which have arithmetic analogues to be discussed below. Our exposition is aimed at making the comparison easy.

Given a manifold M and a vector bundle $E \to M$ of rank l over it, the k-th jet bundle $J^k(E) \to M$ is defined as follows. Let $s \in \Gamma(M, E)$ be a section of the bundle, and $p \in M$. Consider a basis $s_1, \ldots, s_l$ of local sections about p defined on a neighborhood U of a local chart (U, φ), $\varphi(q) = (x_1(q), \ldots, x_n(q))$. We represent s nearby p as $s = f_1 s_1 + \cdots + f_l s_l$ for some smooth functions $f_1, \ldots, f_l$ that are supported in U, and define s to be in $Z_p^k(M, E)$ if, and only if, $(\partial_{x_1}^{\alpha_1} \cdots \partial_{x_n}^{\alpha_n} f_j)(p) = 0$ for all multi-indices $\alpha = (\alpha_1, \ldots, \alpha_n)$ such that $|\alpha| \le k$, for each j. The fiber $J_p^k(E, M)$ of the k-th jet bundle over p is by definition the quotient $\Gamma(M, E)/Z_p^k(M, E)$. If we let $j_k(s)(p)$ denote the equivalence class of s in $J_p^k(E, M)$, $J^k(E, M)$ is given the topology that makes $j_k(s) \in \Gamma(M, J^k(E, M))$ a smooth section.

If E is locally trivialized by $U \times \mathbb{R}^l$, then $J^k(E, M)$ is locally trivialized by $U \times \oplus_{j=0}^k S^j(U, \mathbb{R}^l)$, where $S^j(U, \mathbb{R}^l)$ is the space of symmetric j-linear maps from U into $\mathbb{R}^l$. Let $\{e_1, \ldots, e_n\}$ be the standard basis for $\mathbb{R}^n$, and set $e_i^{\alpha_i}$ to be the α_i-tuple $(e_i, \ldots, e_i)$. Then a local section T of $S^j(U, \mathbb{R}^l)$ can be identified with $\{T(e_1^{\alpha_1}, \ldots, e_n^{\alpha_n})\}_{|\alpha|=j} \in \oplus_{|\alpha|=j} \mathbb{R}^l$, and $j_k(s) \in \Gamma(M, J^k(E, M))$ is then given by the $(k+1)$-tuple of mappings $j_k(s) = (s, Ds, D^2 s, \ldots, D^k s)$, where by definition we have that $(D^{|\alpha|} s)(e_1^{\alpha_1}, \ldots, e_n^{\alpha_n}) = (\partial_{x_1}^{\alpha_1} \cdots \partial_{x_n}^{\alpha_n} s)$, and thus, $j_k(s)$ corresponds to $\{\partial_{x_1}^{\alpha_1} \cdots \partial_{x_n}^{\alpha_n} s\}_{|\alpha| \le k}$ in $\oplus_{|\alpha| \le k} \mathbb{R}^l$.

An operator $F : \Gamma(M, E_1) \to \Gamma(M, E_2)$ between sections of two bundles E_1 and E_2 over M is said to be a differential operator of order k if it factors through the jet bundle $J^k(E_1, M)$ over M, that is to say, if for any $s \in \Gamma(M, E_1)$, we have that

$$F(s)(x) = F(j_k(s))(x) = F(x, s(x), (Ds)(x), (D^2 s)(x), \ldots, (D^k s)(x)).$$

We shall see later in this chapter that arithmetic differential operators can be introduced in a manner that parallels this.

In the arithmetic theory of differential operators, all the concepts above have analogues. At the very elementary level, the role of the manifold M can be taken to be now that of the scheme $\operatorname{Spec} A$, where $A = \mathbb{Z}[q]$, the analogue of the total space E of a bundle $E \to M$ that of a scheme X of finite type over A, and the analogue of the space of sections $\Gamma(M, E)$ that of the set $X(A)$ of A-points of the scheme X. Suitable arithmetic jet spaces can be defined, and with them in place, the arithmetic differential operators can be presented as mappings that factor through them, in exactly the same way as the one described above for the standard differential operators on the manifold M.

6.1 Basic algebraic concepts

Since we have arrived at the study of p-adic analysis and arithmetic differential operators from the point of view of the classical analyst, we attempt to make the presentation self-contained for all readers by briefly recalling the standard concepts in algebraic geometry and commutative algebra that we shall use. The reader with a strong background in algebra may want to bypass this section altogether. Our basic references for the concepts recalled here are [1] and [26], respectively.

Let A be a commutative ring with unity. Given a prime ideal $\mathfrak{p}$ in A, we let S be the multiplicatively closed set $A \setminus \mathfrak{p}$. Then the *localization* $A_\mathfrak{p}$ of A at $\mathfrak{p}$ is the ring of fractions $S^{-1}A$, where a/b and c/d are considered equivalent if $v(ad - bc) = 0$ for some $v \in S$. Notice that the elements of S become invertible in $A_\mathfrak{p}$, ring that only encodes information nearby $\mathfrak{p}$.

Let $f \in A$ and $S = \{f^n\}_{n \geq 0}$. In this case, it customary to write A_f for $S^{-1}A$.

The mapping $a \mapsto a/1$ defines a ring homomorphism $A \to S^{-1}A$. Notice that here we make use of the unit of the ring A in an essential manner. Given an ideal $\mathfrak{i}$ in A we denote by $\mathfrak{i}_\mathfrak{p}$ the ideal generated by its image in $A_\mathfrak{p}$. This establishes a 1-to-1 correspondence between the set of prime ideals in A that are contained in $\mathfrak{p}$ and the prime ideals of $A_\mathfrak{p}$.

A *local ring* is a ring with a unique maximal ideal. The ring $A_\mathfrak{p}$ is local, with $\mathfrak{p}_\mathfrak{p}$ being its unique maximal ideal.

Example 6.1 If $A = \mathbb{Z}$ and $\mathfrak{p} = (p)$ is the ideal generated by a prime p, then $A_\mathfrak{p} = \{m/n : m, n \in \mathbb{Z}, (n, p) = 1\}$. If $f \in A$, $A_f = \{m/f^n : m, n \in \mathbb{Z}, n \geq 0\}$. $\square$

Example 6.2 Let A be the polynomial ring $A = k[t]$ where k is a field and t is an n-tuple of indeterminates. If $\mathfrak{p}$ is a prime ideal of A, the localization $A_\mathfrak{p}$ is the ring of all rational functions $f(t)/g(t)$ where $g \notin \mathfrak{p}$. Suppose that k is algebraically closed, and let $V = V(\mathfrak{p}) = \{x \in k^n : f(x) = 0 \text{ for all } f \in \mathfrak{p}\}$, the *variety* defined by $\mathfrak{p}$. Then $A_\mathfrak{p}$ can be identified with the ring of rational functions $f(x)/g(x)$ in k^n such that $g(x) \neq 0$ at at least one point x in V. $\square$

The prime spectrum $\operatorname{Spec} A$ of a ring A and its Zariski topology are among the most important concepts in algebraic geometry. As a set, $\operatorname{Spec} A$ is the set of all prime ideals of A. Given an ideal $\mathfrak{a}$ in A, we define $V(\mathfrak{a})$ to be the set of all prime ideals containing $\mathfrak{a}$. With more generality, given $E \subset A$, let $V(E)$ denote the set of all primes ideals containing E. We have that $V(E) = V(\mathfrak{a}_E)$ where $\mathfrak{a}_E$ is the ideal generated by E. We obtain a topology on $\operatorname{Spec} A$ by declaring its closed sets to be all sets of the form $V(E) = V(\mathfrak{a}_E)$ for $E \subset A$. This is the *Zariski topology* on $\operatorname{Spec} A$.

Example 6.3   1. For the ring $A = \mathbb{Z}$, we have that

$$\operatorname{Spec} A = \{(0)\} \cup \{(p)\,|\,p \text{ a prime number}\}\,.$$

2. Let k be algebraically closed, and consider the polynomial ring in two variables $x = (x_1, x_2)$, $A = k[x] = k[x_1, x_2]$. Then

$$\operatorname{Spec} A = \{(0)\} \cup \{(p)\,|\,p(x) \text{ irreducible}\} \cup \{(x_1 - a, x_2 - b)\,|\,(a, b) \in k^2\}\,.$$

$\square$

In the case of a multiplicative closed set S, the homomorphism

$$\varphi : A \to S^{-1}A$$

induces an associated mapping ${}^a\varphi : \operatorname{Spec}(S^{-1}A) \to \operatorname{Spec} A$ that is injective. The image $U_S = {}^a\varphi(\operatorname{Spec}(S^{-1}A)) \subset \operatorname{Spec} A$, which consist of the set of prime ideals of A that are disjoint from S, is provided with the subspace topology, and the inverse map $\psi : U_S \to \operatorname{Spec}(S^{-1}A)$ is continuous. Thus, $\operatorname{Spec}(S^{-1}A)$ is homeomorphic to $U_S \subset \operatorname{Spec} A$. When $S = \{f^n : n \geq 0\}$ for f in A, then U_S is the open set $U_S = \operatorname{Spec}(A\backslash V(f))$. These sets are called *principal open sets*, and are denoted it by $D(f)$. They form a basis for the open sets of the Zariski topology.

The sheaf $\mathcal{O}_{\operatorname{Spec} A}$ on $\operatorname{Spec} A$ is defined as follows. Given any open set $U \subset \operatorname{Spec} A$, $\mathcal{O}(U)$ consists of all functions $s : U \to \prod_{\mathfrak{p}\in U} A_\mathfrak{p}$ such that for each $\mathfrak{p} \in U$ we have that $s(\mathfrak{p}) \in A_\mathfrak{p}$, and there exists a neighborhood $V \subset U$ of $\mathfrak{p}$, and elements a, b of A such that, for each $\mathfrak{q} \in V$, $b \notin \mathfrak{q}$, and $s(\mathfrak{q}) = a/b$ in $A_\mathfrak{q}$. The sections so defined form a sheaf $\mathcal{O}_{\operatorname{Spec} A}$ with the induced sum and product operations.

The *spectrum* of A is the topological space $\operatorname{Spec} A$ together with the sheaf $\mathcal{O}_{\operatorname{Spec} A}$ on $\operatorname{Spec} A$.

A *ringed space* is a pair $(X, \mathcal{O}_X)$ consisting of a topological space X and a sheaf of rings $\mathcal{O}_X$ on X. If $\mathcal{F}$ is a sheaf of rings on a topological space X, and $p \in X$, the stalk $\mathcal{F}_p$ at p is defined to be the *germ* of sections at p, sections defined in a neighborhood of p where any two such are considered equivalent if their restrictions agree on a smaller neighborhood that contains p. The stalk $\mathcal{F}_p$ is a ring. A morphism of ringed spaces $\varphi : (X, \mathcal{O}_X) \to (Y, \mathcal{O}_Y)$ is given by a continuous mapping $X \xrightarrow{\varphi} Y$ and a collection of homomorphism $\psi_U : \mathcal{O}_Y(U) \to \mathcal{O}_X(\varphi^{-1}(U))$, U open in Y such that $\psi_U \circ \rho_{\varphi^{-1}(U),\varphi^{-1}(V)} = \rho_{U,V} \circ \psi_V$ for any open sets U, V, $U \subset V$, $\rho_{U,V}$ the restriction homomorphism of the sheaf. An isomorphism between ringed spaces is a morphism of ringed spaces that has an inverse.

A *locally ringed space* $(X, \mathcal{O}_X)$ is a topological space X with a sheaf $\mathcal{O}_X$ on X such that the stalk $\mathcal{O}_{X,p}$ is a local ring for each point $p \in X$. A morphism $\varphi : (X, \mathcal{O}_X) \to (Y, \mathcal{O}_Y)$ of locally ringed spaces is a morphism of ringed spaces that induces a local homomorphism between the stalks of $\mathcal{O}_Y$ and the stalks of $\mathcal{O}_X$, a homomorhism being local if for every $x \in X$ the maximal ideal of the local ring or stalk at $\varphi(x) \in Y$ is mapped to the maximal ideal of the local ring at $x \in X$.

An *affine scheme* is a locally ringed space $(X, \mathcal{O}_X)$ that is isomorphic as a locally ringed space to the spectrum of some ring. A *scheme* is a locally ringed space $(X, \mathcal{O}_X)$ with the property that every point $p \in X$ has a neighborhood U so that the ringed space $(U, \mathcal{O}_U)$ is isomorphic to $(\operatorname{Spec} A, \mathcal{O}_{\operatorname{Spec} A})$ for some ring A.

A morphism of schemes is a morphisms of locally ringed spaces, and an isomorphism is a morphism with a two-sided inverse.

Given a ring R, the *affine n-space* $\mathbb{A}^n$ over R is defined to be the locally ringed space $(\operatorname{Spec} R[x^1, \dots, x^n], \mathcal{O}_{\operatorname{Spec} R[x^1,\dots,x^n]})$. This notion coincides with the earlier one introduced in §3.3.

Example 6.4 Let A be a ring. The subspace of $\operatorname{Spec} A$ consisting of all maximal ideals of A with the induced topology is the *maximal spectrum* of A, and is denoted by $\operatorname{Max} A$. Let $\mathcal{O}_{\operatorname{Max} A}$ be the sheaf of rings in this case. Since the inverse image of a maximal ideal need not be maximal, $\operatorname{Max} A$ does not have the same nice functorial properties of $\operatorname{Spec} A$.

This algebraic set-up ties well with classical analysis. For we may realize a paracompact Hausdorff space X in terms of its ring of continuous functions $C(X)$. Indeed, for each $x \in X$, we may define

$$\mathfrak{m}_x = \{f \in C(X) : f(x) = 0\},$$

a maximal ideal. If $\tilde{X} = \operatorname{Max} C(X)$, we obtain a mapping

$$X \to \tilde{X}$$
$$x \mapsto \mathfrak{m}_x$$

By Urysohn's lemma, this map is injective, and by a partition of unity argument it can be proven to be surjective if we assume X to be compact. In this case, since the open sets $U_f = \{x \in X : f(x) \neq 0\}$ and $\tilde{U}_f = \{\mathfrak{m} \in \tilde{X} : f \notin \mathfrak{m}\}$ form bases for the topologies of X and $\tilde{X}$, respectively, it is easy to see that this map is a homeomorphism. $\quad\square$

If $\mathbb{K}$ is an algebraically closed field, the set of maximal ideals in $A = \mathbb{K}[x^1, \ldots, x^n]$ corresponds to the set of n-tuples $(k^1, \ldots, k^n) \in \mathbb{K}^n$, as a maximal ideal $\mathfrak{m}$ is of the form $\mathfrak{m} = (x^1 - k^1, \ldots, x^n - k^n)$ for some $k^1, \ldots, k^n \in \mathbb{K}$. With the induced topology, $\mathrm{Max}\, A$ is identified with $\mathbb{K}^n$. An *algebraic variety* X over $\mathbb{K}$ is a locally ringed space that is locally isomorphic to $(\mathrm{Max}\, A, \mathcal{O}_{\mathrm{Max}\, A})$, where A is a finitely generated algebra over $\mathbb{K}$. An algebraic variety is an affine variety if it is isomorphic to $(\mathrm{Max}\, A, \mathcal{O}_{\mathrm{Max}\, A})$.

Schemes are the rightful generalizations of algebraic varieties. They parallel the idea of a manifold in classical analysis or geometry, in particular when we think of reconstructing a paracompact space by its ring of continuous functions, as in Example 6.4. Schemes allow for viewing algebraic varieties invariantly, capturing their essential algebraic properties, and as such, lead to other ideas and methods in their study, extending the range of applicability of the concept.

A Noetherian ring is a ring where every ideal is finitely generated. A scheme $(X, \mathcal{O}_X)$ is locally Noetherian if it can be covered by open affine subsets $\mathrm{Spec}\, A_i$ where each A_i is a Noetherian ring. The scheme $(X, \mathcal{O}_X)$ is said to be Noetherian if it can be covered by a finite number of open affine subsets $\mathrm{Spec}\, A_i$ where each A_i is a Noetherian ring.

Let S be a fixed scheme. A scheme X over S is a scheme $(X, \mathcal{O}_X)$ together with a morphism $X \to S$. A morphism $f : X \to Y$ of schemes X, Y over S is a morphism compatible with the morphisms to S from X and Y, respectively. A scheme X over a ring A is a scheme $(X, \mathcal{O}_X)$ together with a morphism $(X, \mathcal{O}_X) \to (\mathrm{Spec}\, A, \mathcal{O}_{\mathrm{Spec}\, A})$.

Given a scheme X over A, the set $X(A)$ of A-points of X are the sections of $X \to \mathrm{Spec}\, A$, that is to say, $\mathrm{Hom}_A(\mathrm{Spec}\, A, X)$.

A scheme X over a ring A is of finite type if X admits a finite covering of the form $X = \cup U_i$, where $U_i = \mathrm{Spec}\, A_i$ and A_i is a ring that is a finitely generated algebra over A.

Remark 6.5 Let $\mathcal{S} = \{schemes\}$ be the category of schemes, and let $\mathcal{F} = \{\text{functors } \{rings\} \to \{sets\}\}$. Schemes can be reconstructed up to isomorphism from the functor

$$\varphi : \mathcal{S} \to \mathcal{F},$$

obtained by defining $\varphi(X)$ to be the functor that associates to a ring B the set of B-points $X(B)$ of X, and by sending a morphism $X \xrightarrow{f} Y$ to the mapping $X(B) \xrightarrow{f} Y(B)$ that takes $u \in \mathrm{Hom}(\mathrm{Spec}\, B, X)$ into $f \circ u \in \mathrm{Hom}(\mathrm{Spec}\, B, Y)$. For instance, if we consider the affine space $X = \mathrm{Spec}\, A[x, y] = \mathbb{A}^2$ over A, then its set of B points is $\mathbb{A}^2(B) = B \times B$, while for the affine scheme $X = \mathrm{Spec}(A[x, y]/f(x, y))$ the set of B points is $X(B) = \{(b_1, b_2) \in B \times B : f(b_1, b_2) = 0\}$.

A scheme X can be reconstructed up to isomorphism by its image under φ.

If the ground ring is algebraically closed, we can consider the category $\mathcal{V}$ of varieties and morphisms. We can then define a functor

$$\overline{\varphi} : \mathcal{V} \to \mathcal{S}$$

that identifies a variety with the corresponding scheme, and a morphism between varieties with the corresponding morphism of schemes. $\square$

We shall need the concept of completion of a ring A at an ideal, which geometrically concentrates attention on a formal neighborhood of a point, or on a Zariski closed subscheme of its spectrum $\mathrm{Spec}\, A$.

For if $\mathfrak{i}$ is an ideal of the ring A, then it induces a topology on A where a basis of open neighborhoods of 0 is given by the nested sequence of powers

$$A \supset \mathfrak{i} \supset \mathfrak{i}^2 \supset \cdots \supset \mathfrak{i}^n \supset \cdots.$$

If $A_n = A/\mathfrak{i}^n$ and $\varphi_n : A_n \to A_{n-1}$ is the natural mapping, we obtain a sequence

$$\cdots \to A_n \xrightarrow{\varphi_n} A_{n-1} \xrightarrow{\varphi_{n-1}} \cdots \to A_2 \xrightarrow{\varphi_2} A_1.$$

The completion $\widehat{A^{\mathfrak{i}}}$ is the inverse limit of this sequence as n goes to infinity: $\widehat{A^{\mathfrak{i}}} = \varprojlim(A_n, \varphi_n)$. It is a complete topological ring.

Example 6.6 1. The p-adic integers $\mathbb{Z}_p$, as constructed in (2.7), are the completion of $\mathbb{Z}$ at the ideal (p) generated by p. The completion of $\mathbb{Z}_p^{ur}$ at (p) is the basic ring $\widehat{\mathbb{Z}}_p^{ur}$ of Buium's theory.

2. The completion of a polynomial ring $A[x]$ at the ideal generated by x is the ring $A[[x]]$ of formal power series. $\square$

If A is a Noetherian ring, its completion $\widehat{A^i}$ at i is again Noetherian. We obtain a canonical continuous homomorphism $i : A \to \widehat{A^i}$ whose kernel consists of the zero divisors a such that $a - 1 \in i$. If the homomorphism i is an isomorphism, then the ring A is said to be complete with respect to i. The ring $\widehat{A^i}$ is complete with respect to $i(i)\widehat{A^i} \subset \widehat{A^i}$.

Let A be a Noetherian ring that is complete with respect to i. On $X = V(i) \subset \operatorname{Spec} A$ we define the sheaf of topological rings $\mathcal{O}_X$ by $\Gamma(\mathcal{D}(f), \mathcal{O}_X) = \varprojlim A_f / i^n A_f$ for $\mathcal{D}(f) = D(f) \cap X$. We call the pair $(X, \mathcal{O}_X)$ the formal spectrum of A, and denote it by $\operatorname{Spf}(A)$, referring to i as the defining ideal of $\operatorname{Spf}(A)$. A formal scheme is a topological local ringed space that is locally isomorphic to a formal spectrum.

In a ring A, a finite strictly increasing chain of prime ideals $\mathfrak{p}_0 \subset \mathfrak{p}_1 \subset \cdots \subset \mathfrak{p}_n$ is said to have length n. The dimension of A is the supremum of the lengths of all chains of prime ideals. For instance, $\dim \mathbb{Z} = 1$. In a Noetherian local ring A of dimension n, let $\mathfrak{m}$ be its maximal ideal. Then $\mathfrak{m}/\mathfrak{m}^2$ has the structure of a vector space over the field $k = A/\mathfrak{m}$. The Noetherian ring is said to be regular if $\dim_k \mathfrak{m}/\mathfrak{m}^2 = n$. A scheme is said to be regular if all of its local rings are regular local rings. A scheme X of finite type over a field k is said to be a smooth scheme if the scheme obtained from X by pullback from the field k to its algebraic closure $\overline{k}$ is a regular scheme. These two notions are closely related to each other, and coincide if the field k is perfect. In particular, a smooth scheme of finite type over an algebraically closed field k is a nonsingular algebraic variety.

Example 6.7   We consider the classical notion of an affine variety X over $\mathbb{C}$ from the algebraic point of view summarized above.

Let X be the maximal spectrum of some finitely generated $\mathbb{C}$-algebra A. We then have $A = \mathbb{C}^n/\mathfrak{p}$ for some ideal $\mathfrak{p}$. The maximal spectrum $\operatorname{Max} \mathbb{C}[x^1, \ldots, x^n]$ of the polynomial algebra $\mathbb{C}[x^1, \ldots, x^n]$ identifies with $\mathbb{C}^n$, and by Hilbert's basis theorem, the ring $\mathbb{C}[x^1, \ldots, x^n]$ is Noetherian. Thus, the ideal $\mathfrak{p}$ viewed in the polynomial algebra is generated by finitely many polynomials $p_1, \ldots, p_k$, and the affine variety X becomes identified with the maximal spectrum of $\mathbb{C}[x^1, \ldots, x^n]/(p_1, \ldots, p_k)$, the ring of functions on X.

On the other hand, if we start with the ring of functions $\mathcal{O}_X$, we can take a finite number of generators $x^1, \ldots, x^n$, and using the Hilbert's basis theorem, find a finite number of relations $p_1, \ldots, p_k$ among them. Then we can set X to be the affine variety in $\mathbb{C}^n$ cut out by the polynomials $p_1, \ldots, p_k$, $X = \{x \in \mathbb{C}^n : p_1(x) = \cdots = p_k(x) = 0\}$.

The embedding of the affine variety X into $\mathbb{C}^n$ corresponds to a choice of generators $x^1, \ldots, x^n$ of $\mathcal{O}_X$. Each generator yields a map $X \to \mathbb{C}$, and all of them together the map $X \hookrightarrow \mathbb{C}^n$. This may be stated in an invariant manner. Indeed, we consider the embedding into the dual of the vector space $\langle x^1, \ldots, x^n \rangle_{\mathbb{C}}$ spanned by the generators, where $x \in X$ gets identified with the evaluation at x mapping $\mathrm{ev}_x : \langle x^1, \ldots, x^n \rangle_{\mathbb{C}}^*$ that carries $f \in \mathbb{C}[x^1, \ldots, x^n]$ into $\mathrm{ev}_x f = f(x)$. If the variety X is irreducible, the ideal of functions $(p_1, \ldots, p_k)$ vanishing on $X \subset \mathbb{C}^n$ is prime. This is reflected in the fact that $\mathcal{O}_X$ contains no zero divisors. By definition, a variety is irreducible if it cannot be written as the union of two nonempty closed subsets.

The coordinate free approach follows the ideas in Example 6.4. The points x of an affine variety X are in one-to-one correspondence with the maximal ideals $\mathfrak{i}_x \subset \mathcal{O}_X$ of functions vanishing at x. So if A is a ring, we associate with it the affine variety $X = \mathrm{Spec}\, A$ whose points are the maximal ideals in A. We get the identifications $\mathcal{O}_X = A$ and $X = \mathrm{Spec}\, A$, respectively. If the ring A has no zero divisors, the variety X is irreducible.

The identification outlined can be extended to arbitrary affine schemes, allowing nilpotents in their ring of functions. These would correspond to multiplicities, or infinitesimal directions, in the scheme.   □

6.2 General δ_p-functions: arithmetic jet spaces

We use the remaining parts of this chapter to provide the reader the announced fuller view of arithmetic differential operators, and their general theory.

For simplicity, we set $R = \widehat{\mathbb{Z}}_p^{ur}$, and let $k = R/pR$ be the residue field. This ring carries a δ_p-derivation in a natural way, as we will see below, and it is the ground ring of the general theory of arithmetic differential operators. It is a much richer ring than $\mathbb{Z}_p$ in that both, R and k, have very special properties, with the first being complete and the latter being algebraically closed. The theory of arithmetic differential operators is built for any δ_p-ring A over R, that is to say, a δ_p-ring A together with a δ_p-ring homomorphism $R \to A$. In what follows, A will stand for any ring of this type unless otherwise indicated.

On a smooth scheme X over R, the ring of arithmetic differential operators of order r identifies with the ring of global functions of certain formal scheme $J^r(X)$, called the arithmetic r-th p-jet space of X [8]. We begin with the discussion and definition of $J^r(X)$.

Let $s_1(x_0, y_0, x_1, y_1)$ and $s_2(x_0, y_0, x_1, y_1)$ be two elements of the polynomial ring $\mathbb{Z}[x_0, y_0, x_1, y_1]$. Paraphrasing the definition of δ_p-operator given earlier, we say that a mapping

$$\delta : A \to B$$

is a δ-operator over an A-algebra B associated to the pair of polynomials $\{s_1, s_2\}$ if

$$\begin{aligned} \delta(x+y) &= s_1(x, y, \delta x, \delta y)\,, \text{ and} \\ \delta(xy) &= s_2(x, y, \delta x, \delta y)\,. \end{aligned}$$

The pair of polynomials $\{s_1, s_2\} \subset \mathbb{Z}[x_0, y_0, x_1, y_1]$ is said to be δ-generic if for a ring A of characteristic zero there exists an operator $\delta : A \to A$ such that, if $g \in A[x_0, x_1]$ has the property that $g(a, \delta a) = 0$ for all $a \in A$, then $g = 0$.

Two δ-operators $\delta_1 : A \to B$ and $\delta_2 : A \to B$ are said to be equivalent over a subring A_0 of A if there exists a constant $a_0 \in A_0^\times$ and $f \in A_0[x]$ such that $\delta_1 a = a_0 \delta_2 a + f(a)$ for all $a \in A$.

We have the following [5] result, characterizing equivalent δ-operators for generic polynomials:

Theorem 6.8 *Let $\{s_1, s_2\} \subset \mathbb{Z}[x_0, y_0, x_1, y_1]$ be a generic pair over a ring A of characteristic zero, and let $\delta : A \to B$ be a δ-operator associated to this pair. Assume that the localization $\mathbb{Z}_{(p)}$ is contained in A for some fixed prime p. Then δ is equivalent over $\mathbb{Z}_{(p)}$ to the δ-operator associated with one of the following four pairs:*

1. $s_1(x_0, y_0, x_1, y_1) = x_1 + y_1$, $s_2(x_0, y_0, x_1, y_1) = x_0 y_1 + y_0 x_1$,
2. $s_1(x_0, y_0, x_1, y_1) = x_1 + y_1$, $s_2(x_0, y_0, x_1, y_1) = x_0 y_1 + y_0 x_1 + x_1 y_1$,
3. $s_1(x_0, y_0, x_1, y_1) = x_1 + y_1$, $s_2(x_0, y_0, x_1, y_1) = x_0 y_1 + y_0 x_1 + p^n x_1 y_1$, *where n is some integer in $\mathbb{N}$, and*
4. $s_1(x_0, y_0, x_1, y_1) = x_1 + y_1 + (x_0^{p^n} + y_0^{p^n} - (x_0 + y_0)^{p^n})/p$, $s_2(x_0, y_0, x_1, y_1) = x_0^{p^n} y_1 + y_0^{p^n} x_1 + p x_1 y_1$, *where n is some integer in $\mathbb{N}$.*

$\square$

Notice that the δ-operator associated with the first three generic pairs in the theorem above restricts to the zero map on $\mathbb{Z}$, while the δ-operator associated to the last pair yields

$$\delta(m) = \frac{m - m^{p^n}}{p}\,,$$

which coincides with the restriction of (5.5) to $\mathbb{Z}$ in the case when $n = 1$.

We now define the basic δ_p-ring structure on R, as follows.

Let $\mu(R)$ be the multiplicative group of roots of unity in R. The reduction mod p mapping $\mu(R) \to k^\times$ defines an isomorphism whose inverse is the Teichmüller lift. By our discussion in §3.2, any $x \in R$ can be written uniquely as a series

$$x = \sum_{i=0}^{\infty} \zeta_i p^i ,$$

where $\zeta_i \in \mu(R) \cup \{0\}$. We thus obtain a mapping

$$\phi_p : R \to R \tag{6.2}$$

defined by

$$\phi_p \left(\sum_{n=0}^{\infty} \zeta_n p^n \right) = \sum_{n=0}^{\infty} \zeta_n^p p^n .$$

Clearly $\phi_p(\zeta) = \zeta^p$ for $\zeta \in \mu(R)$.

Theorem 6.9 *The mapping* (6.2) *above is a ring homomorphism that lifts the p-th power Frobenius homomorphism on* k.

Proof. Let N be a positive integer coprime to p, and let ζ_N be a primitive N-root of unity. The Galois group $G(\mathbb{Q}(\zeta_N)/\mathbb{Q})$ is naturally isomorphic to $(\mathbb{Z}/N\mathbb{Z})^\times$, with the Galois element s corresponding to the class of the integer n such that $s\zeta_N = \zeta_N^n$ (see discussion in Example 5.8).

We consider the class of p in $(\mathbb{Z}/N\mathbb{Z})^\times$. It corresponds to an automorphism $s = s_p$ of $\mathbb{Q}(\zeta_N)$ such that $s\zeta_N = \zeta_N^p$. The automorphism s induces an automorphism of $\mathbb{Z}[\zeta_N]$, and therefore, an automorphism of the completion A_N of this ring at any prime ideal P_N containing p. In this manner we obtain an induced automorphism $s = s_p$ of the P-adic completion A of the direct limit of all of these A_Ns, where $P \subset A$ is the direct limit of a sequence of ideals P_N. Since A is isomorphic to R, and clearly the automorphism s sends p to p and any ζ_N into ζ_N^p, we conclude that s must be equal to ϕ_p, and so ϕ_p is a ring homomorphism. $\square$

Since δ_p-operators on a δ_p-ring A are in 1-to-1 correspondence with homomorphisms of the ring A (see (5.4)), we may associate to the homomorphism ϕ_p the p-derivation on R given by

$$\delta_p x = \frac{\phi_p(x) - x^p}{p} . \tag{6.3}$$

The pair (R, δ_p) constitutes the basic δ_p-ring of the general theory of arithmetic differential operators.

Let us consider an N-tuple of indeterminates $y = (y_1, \ldots, y_N)$ over R, and let $y^{(i)}$ be a family of N-tuple of indeterminates over R parametrized by $\mathbb{Z}_{\geq 0}$, with $y^{(0)} = y$. We set

$$R\{y\} := R[y^{(i)} \mid_{i \geq 0}]$$

for the polynomial ring in the indeterminates $y^{(i)}$. This ring has a natural δ_p-structure on it.

Proposition 6.10   *There exists a unique p-derivation*

$$\delta_p : R\{y\} \to R\{y\}$$

that extends the p-derivation (6.3) *on R, and satisfies the relations* $\delta_p y_k^{(i)} = y_k^{(i+1)}$.

Proof. This follows by a rather explicit construction. For let

$$\phi_p : R\{y\} \to R\{y\}$$

be the unique ring homomorphism extending the homomorphism (6.2) of Theorem 6.9, and satisfying the relation

$$\phi_p(y^{(i)}) = (y^{(i)})^p + py^{(i+1)}.$$

Then, we may define a p-derivation $\delta_p : R\{y\} \to R\{y\}$ by the expression

$$\delta_p P := \frac{\phi_p(P) - P^p}{p},$$

where

$$\phi_p(P(x^{(0)}, x^{(1)}, x^{(2)}, \ldots)) = P^{\phi_p}((x^{(0)})^p + px^{(1)}, (x^{(1)})^p + px^{(2)}, \ldots).$$

Here, P^{ϕ_p} stands for the polynomial obtained from P by twisting its coefficients by ϕ_p, and $(x^{(i)})^p$ is the tuple of p-th powers of the components of $x^{(i)}$.

This definition implies that $\delta_p y^{(i)} = y^{(i+1)}$, as desired. $\square$

We now take advantage of the fact that R is complete, and that the residue field $k = R/pR$ is algebraically closed, to develop arithmetic analogues of the jet spaces that were outlined above. In the sequel, for a given smooth or formal scheme X over R, we shall denote its ring of sections by $\mathcal{O}(X)$.

Definition 6.11 Let $S^* = \{S^n\}_{n \geq 0}$ be a sequence of rings. Suppose we have ring homomorphisms $\varphi : S^n \to S^{n+1}$, and p-derivations $\delta_p : S^n \to S^{n+1}$ such that $\delta_p \circ \varphi = \varphi \circ \delta_p$. We then say that (S^*, φ, δ_p), or simply S^*, is a δ_p-*prolongation sequence*. A *morphism* $u^* : S^* \to \tilde{S}^*$ of δ_p-prolongation sequences is a sequence $u^n : S^n \to \tilde{S}^n$ of ring homomorphisms such that $\delta_p \circ u^n = u^{n+1} \circ \delta_p$ and $\varphi \circ u^n = u^{n+1} \circ \varphi$, respectively. $\qquad\qquad\square$

If A is any δ_p-ring over R, we obtain a natural δ_p-prolongation sequence A^* by setting $A^n = A$ for all n, and taking the ring homomorphisms φ to be all equal to the identity. A δ_p-*prolongation sequence* over A is a δ_p-prolongation sequence S^* equipped with a morphism $A^* \to S^*$ of prolongation sequences.

For the ring $R\{y\}$ discussed above, we consider the subrings

$$S_R^n := R[y, \delta_p y, \dots, \delta_p^n y].$$

We view S_R^{n+1} as an S_R^n-algebra via the inclusion homomorphism, and we observe that $\delta_p S_R^n \subset S_R^{n+1}$. Therefore, the sequence of rings $S_R^* = \{S_R^n\}$ defines a δ_p-prolongation sequence. Let (p) be the ideal generated by p. We obtain the p-adic completion prolongation sequence

$$R[y, \delta_p y, \dots, \delta_p^n y]^{\widehat{(p)}},$$

and the prolongation sequence

$$R[[y, \delta_p y, \dots, \delta_p^n y]]$$

of formal power series rings, with their corresponding δ_p-structures.

We now consider a scheme X of finite type over R, and define its δ_p-*jet spaces*. We begin with the case where X is an affine scheme of finite type, that is to say, the spectrum of a finitely generated R-algebra $X = \operatorname{Spec}(R[y]/I)$, where y is a tuple of indeterminates, and I is some ideal. Then we set

$$J_p^n(X) := \operatorname{Spf} R[y, \delta_p y, \dots, \delta_p^n y]^{\widehat{(p)}}/(I, \delta_p I, \dots, \delta_p^n I). \qquad (6.4)$$

Notice that $J_p^0(X) = X^{\widehat{(p)}}$, and that the sequence of rings of sections given by $\{\mathcal{O}(J_p^n(X))\}_{n \geq 0}$ has a natural structure of δ_p-prolongation sequence $\mathcal{O}(J_p^*(X))$ induced by the p-derivation δ_p in Proposition 6.10.

The δ_p-prolongation sequence so constructed has the following universality property.

Proposition 6.12 Let X be an affine scheme over R. Then for any δ_p-prolongation sequence S^* over R and any homomorphism $u : \mathcal{O}(X) \to S^0$, there exists a unique morphism of δ_p-prolongation sequences

$$u_* = (u_n) : \mathcal{O}(J_p^*(X)) \to S^*,$$

such that $u_0 = u$.

Proof. It suffices to consider the case where $X = \operatorname{Spec}(R[y]/I)$ with $I = \{0\}$. We define

$$u_n : R[y^{(0)}, \ldots, y^{(r)}]^{\widehat{(p)}} \to S^r$$

by

$$u_n(y^{(i)}) = \delta_p^i(u(y^{(0)})).$$

This defines a morphism of prolongation sequences. This morphism has the desired properties. $\square$

The construction $X \to J_p^r(X)$ above for affine schemes over R is compatible with localization in the following sense:

Corollary If $X = \operatorname{Spec} B$ and $U = \operatorname{Spec} B_f$, $f \in B$, then

$$\mathcal{O}(J_p^r(U)) = (\mathcal{O}(J_p^r(X))_f)^{\widehat{(p)}}.$$

Equivalently,

$$J_p^r(U) = J_p^r(X) \times_{X^{\widehat{(p)}}} U^{\widehat{(p)}}.$$

Proof. The rings $(\mathcal{O}(J_p^r(X))_f)^{\widehat{(p)}}$ has the structure of a prolongation sequence by defining

$$
\begin{aligned}
\delta_p\left(\frac{g}{f^n}\right) &= \left(\frac{f^{np}\delta_p g - g^p \delta_p(f^n)}{f^n \phi_p(f^n)}\right) \\
&= \left(\frac{f^{np}\delta_p g - g^p \delta_p(f^n)}{f^{2np}}\right)\left(1 - p\left(\frac{\delta_p f}{f^p}\right) + p^2\left(\frac{\delta_p f}{f^p}\right)^2 \right. \\
&\qquad \left. -p^3\left(\frac{\delta_p f}{f^p}\right)^3 + \cdots\right).
\end{aligned}
$$

This prolongation sequence satisfies the same universality property as that of the rings $\mathcal{O}(J_p^r(U))$. $\square$

Definition 6.13 Let X be a scheme of finite type over R. For any affine Zariski open covering of X

$$X = \bigcup_{i=1}^m U_i$$

we define the δ_p-*jet space of order r* of X by gluing $J_p^r(U_i)$ along $J_p^r(U_i \cap U_j)$. We denote the resulting formal schemes by $J_p^r(X)$, and refer to it as the r-th p-jet space of X. $\Box$

The resulting formal schemes have a universality property whose formulation is parallel to the one in Proposition 6.12. Using this property we see that, up to isomorphism, the formal schemes $J_p^r(X)$ depend only on X in a functorial manner: for any morphism $X \to Y$ of affine schemes of finite type, there exist induced morphisms of schemes

$$J_p^r(Y) \to J_p^r(X).$$

Notice that if X itself is affine, we may use the *tautological covering* of X consisting of the covering of X given by just one open set, namely X itself. The space $J_p^r(X)$ with respect to this tautological covering coincides with the space $J_p^r(X)$ in (6.4). This justifies the abuse of notation we have incurred into when using the same terminology to denote the jet space of X, whether X is affine or not.

Definition 6.14 Let X be a scheme of finite type over R. The δ_p-*jet spaces* of X are given by the projective system of p-adic formal schemes

$$\cdots \to J_p^r(X) \to J_p^{r-1}(X) \to \cdots \to J_p^1(X) \to J_p^0(X) = X^{\widehat{(p)}}. \quad (6.5)$$

$\Box$

Definition 6.15 Let X and Y be smooth schemes over the fixed δ_p-ring A. By a δ_p-*morphism of order r* we mean a rule $f : X \to Y$ that attaches to any δ_p-prolongation sequence S^* of p-adically complete rings, a map of sets $X(S^0) \to Y(S^r)$ that is "functorial" in S^* in the obvious sense. $\Box$

For any δ_p-prolongation sequence S^*, the shifted sequence $S^*[i]$ defined by $S[i]^n := S^{n+i}$ is a new δ_p-prolongation sequence. Thus, any morphism $f : X \to Y$ of order r induces maps of sets $X(S^i) \to Y(S^{r+i})$ that are functorial in S^*. We can compose δ_p-morphisms $f : X \to Y$, $g : Y \to Z$ of orders r and s, respectively, and get δ_p-morphisms $g \circ f : X \to Z$ of order $r + s$. There is a natural map from the set of δ_p-morphisms $X \to Y$ of order r into the set of δ_p-morphisms $X \to Y$ of order $r + 1$, induced by the maps $Y(S^r) \to Y(S^{r+1})$ arising from the S^r-algebra structure of S^{r+1}.

Remark 6.16   By the universality property of jet spaces, we have that

$$J_p^n(X \times Y) \simeq J_p^n(X) \times J_p^n(Y)$$

where the product on the left hand side is taken in the category of schemes of finite type over R, and the product on the right hand side is taken in the category of formal schemes over R. $\square$

Remark 6.17   By the universality property of jet spaces, the set of order r δ_p-morphisms $X \to Y$ between two schemes of finite type over R naturally identifies with the set of morphisms over R of formal schemes $J_p^r(X) \to J_p^0(Y) = Y^{\widehat{(p)}}$. $\square$

Proposition 6.18   *Let X be a smooth affine scheme over A, and let $u : A[y] \to \mathcal{O}(X)$ be an étale morphism, where y is a N-tuple of indeterminates. Let $y^{(i)}$ be N-tuples of indeterminates parametrized by $i \in \mathbb{Z}_{\geq 0}$. Then the natural morphism*

$$\mathcal{O}(X^{\widehat{(p)}})[y^{(i)} \mid_{1 \leq i \leq n}]^{\widehat{(p)}} \to \mathcal{O}(J_p^n(X))$$

that sends $y^{(i)}$ into $\delta_p^i(u(y))$ is an isomorphism. In particular, we have an isomorphism of formal schemes over A

$$J_p^n(X) \simeq X^{\widehat{(p)}} \times (\mathbb{A}^{nN})^{\widehat{(p)}}.$$

Corollary   *If $Y \to X$ is an étale morphism of smooth schemes over A, then*

$$J_p^n(Y) \simeq J_p^n(X) \times_{X^{\widehat{(p)}}} Y^{\widehat{(p)}}.$$

Remark 6.19   The functors $X \mapsto J_p^r(X)$ from the category $\mathcal{C}$ of A-schemes of finite type to the category $\hat{\mathcal{C}}$ of p-adic formal schemes naturally extends to a functor from $\mathcal{B}$ to $\hat{\mathcal{C}}$, where $\mathcal{B}$ is the category whose objects are the same as those in $\mathcal{C}$, hence $Ob\,\mathcal{C} = Ob\,\mathcal{B}$, and whose morphisms are defined by

$$\mathrm{Hom}_{\mathcal{B}}(X, Y) := \mathrm{Hom}_{\hat{\mathcal{C}}}(X^{\widehat{(p)}}, Y^{\widehat{(p)}})$$

for all $X, Y \in Ob\,\mathcal{B}$. $\square$

We now spell out the relationship between p-jet spaces and *arithmetic differential operators* in this degree of generality. Let X be a scheme of finite type over R. A global function $f \in \mathcal{O}(J_p^r(X))$ is an arithmetic differential equation of order r. It induces a map of sets $f : X(R) \to R$.

We want to cast this definition in a manner reminiscent of that given in Definition 5.3, and reminiscent also of the definition of a classical differential operator recalled at the beginning of this chapter.

For any nonnegative integer r, we obtain a natural map ∇^r between the set of R-points $X(R)$ and $J_p^r(X)(R)$,

$$\nabla^r : X(R) \to J^r(X)(R),$$

and an R-algebra map

$$\begin{array}{ccc} \mathcal{O}(J_p^r(X)) & \to & \mathcal{O}^r(X) \\ f & \mapsto & \tilde{f}. \end{array}$$

In order to see this, and since the assertion is local, it will suffice to consider the affine case where $X = \operatorname{Spec}(R[y]/I)$ for some ideal I. We then recall that an R-point x in $X(R)$ is a morphism of R-schemes $\operatorname{Spec} R \to X$, which induces a homomorphism $\mathcal{O}(X) \to R$ that we denote by x also, abusing the notation some. By the universality property in Proposition 6.12, we obtain a map $\nabla^r(x) : \mathcal{O}(J_p^r(X)) \to R$, which yields an R-point of $J_p^r(X)(R)$. Therefore, given $f \in \mathcal{O}(J_p^r(X))$, we define

$$\tilde{f}(x) = \nabla^r(x)(f).$$

If we are given an embedding $i : \operatorname{Spec}(R[y]/I) \subset \operatorname{Spec} R[y] = \mathbb{A}^N$ into affine space, and take f to be the class of an element

$$F \in R[y^{(0)}, y^{(1)}, \ldots, y^{(r)}]\widehat{^{(p)}},$$

we would obtain that

$$\begin{array}{rcl} \nabla^r(x) & = & (i(x), \delta_p(i(x)), \ldots, \delta_p^r(i(x))), \\ \tilde{f}(x) & = & F(i(x), \delta_p(i(x)), \ldots, \delta_p^r(i(x))), \end{array}$$

which shows that $\tilde{f}$ is a δ_p-function of order at most r. This of course is the content of Definition 5.3, when the role of R is played by the ring $\mathbb{Z}_p$. By gluing over the nonempty intersections of the elements of a covering of a scheme X of finite type over R, we obtain an intrinsic global notion of an arithmetic differential operator.

Let us point out that for an arithmetic differential operator of order r to be defined for a given scheme X of finite type over R, the formal scheme $J_p^r(X)$ must admit globally defined functions.

6.3 The analogue of a δ_p-linear operators for group schemes

A group variety is a variety X together with a morphism $X \times X \to X$ such that the points of X with this operation form a group, and such that the inverse $x \to x^{-1}$ is also a morphism of X. An algebraic group is a synonym for a group variety. If $X \xrightarrow{\beta} S$ is a scheme over S, a *group law* is a morphism $X \times_S S \xrightarrow{\mu} X$. Let $i : X \to X$ and $\varepsilon : S \to X$ be morphisms of schemes such that $\beta \circ \varepsilon = \mathbb{1}_S$, where $\mathbb{1}_S$ stands for the identity morphism of S. Suppose that μ, i and ε satisfy the relations $\mu \circ (\varepsilon \circ \beta, \mathbb{1}_X) = \mu \circ (\mathbb{1}_X, \varepsilon \circ \beta) = \mathbb{1}_X$, $\mu \circ (i, \mathbb{1}_X) = \mu \circ (\mathbb{1}_X, i) = \varepsilon \circ \beta$, and $\mu \circ (\mathbb{1}_X, \mu) = \mu \circ (\mu, \mathbb{1}_X)$, where $(\mathbb{1}_X, \mu)$ and $(\mu, \mathbb{1}_X)$ are the two natural morphisms $X \times_S X \times_S X \to X$ that μ and $\mathbb{1}_X$ define. Then we say that the scheme X over S together with the morphisms μ, i and ε is a *group scheme* over S.

In the general set-up of the previous section, we take X to be a commutative group scheme G over A. An arithmetic differential operator gives rise to a mapping $G(A) \to A$. If this mapping is also a group homomorphism when A is viewed with its additive group structure, we say that the operator is a δ_p-character. These are the analogues of the *linear differential operators* in the arithmetic case, as proposed in [6, 8].

The affine line $\mathbb{A}^1$ under addition is an algebraic group, denoted by $\mathbb{G}_a$. Thus, we have that $\mathbb{G}_a = \operatorname{Spec} R[y]$ is the additive group scheme over R. On the other hand, $\mathbb{A}^1 \setminus \{0\} = \operatorname{Spec} R[y, y^{-1}]$ is a group scheme under multiplication, denoted by $\mathbb{G}_m$. Let us recall that given a scheme X over the ring A, if B is an A-algebra, we let $X(B)$ denote the set of all morphisms of A-schemes $\operatorname{Spec} B \to X$, and any such morphism is called a *B-point* of X. If $X = \mathbb{A}^1 = \mathbb{G}_a = \operatorname{Spec} A[y]$, then $X(B)$ is simply the set B itself because a morphism $\operatorname{Spec} B \to \operatorname{Spec} A[y]$ is the same as a morphism $A[y] \to B$, and the latter is uniquely determined by the image of y in B. If on the other hand, $X = \operatorname{Spec} A[y, y^{-1}] = \mathbb{G}_m = \operatorname{Spec} A[x, y]/(xy - 1)$, then $X(B) = B^\times$ because we have the identification $\operatorname{Hom}_A(A[y, y^{-1}], B) = B^\times$ via the map $f \mapsto f(y)$. Finally, if $X = \operatorname{Spec} A[x, y]/(f(x, y))$, then $X(B) = \{(a, b) \in B^2 : f(a, b) = 0\}$. These three examples cover all the algebraic groups of dimension one, as the latter of them encompasses an affine chart of an elliptic curve (E, A) over A. All the geometric fibers of group schemes of relative dimension one are of this form.

Definition 6.20   Let G and H be smooth group schemes over a

δ_p ring A. We say that $G \to H$ is a δ_p-*homomorphism of order* r if it is a δ_p-morphism of order r such that, for any prolongation sequence S^*, the maps $G(S^0) \to H(S^r)$ are group homomorphisms. A δ_p-*character* of order r of G is a δ_p-homomorphism $G \to \mathbb{G}_a$ of order r, where $\mathbb{G}_a = \mathrm{Spec}\, A[y]$ is the additive group scheme over A. The group of δ_p-characters of order r of G will be denoted by $\mathfrak{X}^r_{\delta_p}(G)$. $\qquad\square$

Example 6.21 Let us consider the δ_p-jet space of the group schemes $\mathbb{A}^1$ and $\mathbb{A}^1 \setminus \{0\}$, respectively. In the first case, we have that $\mathbb{A}^1 = \mathbb{G}_a = \mathrm{Spec}\, A[y]$ is the additive group scheme over A, and

$$J^n_p(\mathbb{G}_a) = \mathrm{Spf}\, \widehat{A[y, \delta_p y, \dots, \delta^n_p y]^{(p)}}\,.$$

In the second case, $\mathbb{A}^1 \setminus \{0\} = \mathbb{G}_m = \mathrm{Spec}\, A[y, y^{-1}]$, and

$$J^n_p(\mathbb{G}_m) = \mathrm{Spf}\, \widehat{A[y, y^{-1}, \delta_p y, \dots, \delta^n_p y]^{(p)}}\,.$$

If on the other hand the scheme X is an elliptic curve over A given on an affine chart by $\mathrm{Spec}\, A[x,y]/f(x,y)$, if we let $x^{(0)}$ denote the tuple (x, y) then

$$J^n_p(E, A) = \mathrm{Spf}\, \widehat{A[x^{(0)}, x^{(1)}, \dots, x^{(n)}]^{(p)}}/(f, \delta_p f, \dots, \delta^n_p f)\,,$$

where $\delta_p f(x^0) = (f((x^{(0)})^p + px^{(1)}) - f^p(x^0))/p$. The entire jet space is obtained by using a covering, and gluing the resulting affine pieces together.

These three cases describe the arithmetic jet spaces of all algebraic groups of dimension one. $\qquad\square$

Notice that when X is a group scheme of finite type over R, then (6.5) is a projective system of groups in the category of p-adic formal schemes over R.

If G is a group scheme of finite type over R, then the group $\mathfrak{X}^r_{\delta_p}(G)$ of order r δ_p-characters $G \to \mathbb{G}_a$ identifies with the group of homomorphisms $J^r_p(G) \to \widehat{\mathbb{G}_a^{(p)}}$, and thus, it identifies with an A-submodule of $\mathcal{O}(J^r_p(G))$. Let

$$\mathcal{O}^\infty_{\delta_p}(G) := \varinjlim \mathcal{O}^r_{\delta_p}(G)$$

be the δ_p-ring of all δ_p-morphisms $G \to \mathbb{A}^1$, and let

$$\mathfrak{X}^\infty_{\delta_p}(G) := \varinjlim \mathfrak{X}^r_{\delta_p}(G)$$

be the group of δ_p-characters $G \to \mathbb{G}_a$. Then $\mathcal{O}_{\delta_p}^\infty(G)$ has a natural structure of $R[\phi_p]$-module, and $\mathcal{X}_{\delta_p}^\infty(G)$ is an $R[\phi_p]$-submodule. Here, $(R[\phi_p], +, \cdot)$ denotes the ring generated by R and the symbol ϕ_p subjected to the relation

$$\phi_p \cdot a = a^{\phi_p} \cdot \phi_p,$$

for all $a \in R$, where $a^{\phi_p} := \phi_p(a)$.

If X is a scheme of finite type over R, and $f \in \mathcal{O}(J_p^r(X))$ is an arithmetic differential equation of order r, let $f : X(R) \to R$ be the corresponding map of sets. If $X = G$ is a group scheme, f is a δ_p-character of order r if the said map of sets is a homomorphism. The pre-image $f^{-1}(0)$ of $0 \in R$ under f is, by definition, the set of solutions of the arithmetic differential equation $f = 0$.

If G is a group scheme of dimension one, the structure of the $R[\phi_p]$-module $\mathcal{O}_{\delta_p}^\infty(G)$ of δ_p-characters has been described by Buium [8] in complete detail. His results are as follows.

Theorem 6.22 [8] *On the additive group $G = \mathbb{G}_a$ over R, any δ_p character f is of the form $f = \sum a_i \phi_p^i$ with $a_i \in R$. That is to say, $\mathcal{O}_{\delta_p}^\infty(G)$ is a free $R[\phi_p]$-module of rank 1 with the identity as generator.*

Proof. Let K be the fraction field $R[1/p]$. Given $f \in \mathcal{X}_{\delta_p}^n(\mathbb{G}_a)$, since $J_p^n(\mathbb{G}_a) = \mathrm{Spf}\, A[y, \delta_p y, \ldots, \delta_p^n y]\widehat{^{(p)}}$, we may identify f with an element $f \in R[[y^{(0)}, y^{(1)}, \ldots, y^{(n)}]]$. We have the mapping

$$\sigma : K[[y_0, y_1, \ldots, y_n]] \to K[[y^{(0)}, y^{(1)}, \ldots, y^{(n)}]]$$

defined by $\sigma(y_0) = y^{(0)}$, $\sigma(y_1) = \phi_p(y^{(0)}) = (y^{(0)})^p + p y^{(1)}$, $\ldots$, $\sigma(y_n) = \phi_p^n(y^{(0)}) = (y^{(0)})^{p^n} + \cdots + p^n y^{(n)}$. This is a K-algebra isomorphism, and $\sigma^{-1} \circ f$ is additive in $y_0, \ldots, y_n$. But the only additive elements in $K[[\phi_p^i|_{0 \le i \le n}]]$ are those in the K-linear span of $\{\phi_p^i|_{0 \le i \le n}\}$. Thus, we can complete the proof by showing that if $\sum_i a_i \phi_p^i \in pR[\phi_p^i|_{0 \le i \le n}]$, we then have $a_i \in pR$, and this is fairly clear. $\square$

Theorem 6.23 [8] *On the multiplicative group $G = \mathbb{G}_m$ over R, there exists a nonzero δ_p-character $f : \mathbb{G}_m(R) = R^\times \to R$ of order 1, unique up to multiplication by a constant, and given by*

$$f_p^1(x) = \sum_{n \ge 1} (-1)^{n-1} \frac{p^{n-1}}{n} \left(\frac{\delta_p x}{x^p} \right)^n,$$

that generates $\mathcal{O}_{\delta_p}^\infty(G)$ as a free $R[\phi_p]$-module.

In light of the more complicated nature of the multiplicative group, the argument to describe its characters is more elaborate than the previous one for $\mathbb{G}_a$.

Proof. Let us first observe that

$$f_p^1(x) = \frac{1}{p} \log\left(\frac{\phi_p(x)}{x^p}\right) = \frac{1}{p} \log\left(1 + p\frac{\delta_p x}{x^p}\right).$$

Since we already know that the n-th jet space of $\mathbb{G}_m$ is given by $J_p^n(\mathbb{G}_m) = $ Spf $A[y, y^{-1}, \delta_p y, \ldots, \delta_p^n y]^{\widehat{(p)}}$, we see easily that f_p^1 is an element of the group $\mathfrak{X}_{\delta_p}^1(\mathbb{G}_m)$, a character of order 1 of $\mathbb{G}_m$.

Given any $f \in \mathfrak{X}_{\delta_p}^n(\mathbb{G}_m)$, we can compose it with the map

$$\mathbb{G}_a^{\widehat{(p)}} \quad \to \quad \mathbb{G}_m^{\widehat{(p)}}$$

$$x \quad \mapsto \quad \exp{(px)}$$

to obtain an additive character $\tilde{f}$, which is then an element of the free $R[\phi_p]$-module in Theorem 6.22. Now observe that if $y = 1 + T$ and $\log y = l(T) = T - T^2/2 + T^3/3 + \cdots$, then

$$f_p^1 = \frac{1}{p}(\phi_p - p)l(T) \in R[[T]][T^{(1)}]^{\widehat{(p)}},$$

and so

$$(\phi_p^i f_p^1) \circ \exp{(pT)} = \phi_p^i\left(\frac{1}{p}(\phi_p - p)\right) l(\exp{(pT)}) = \phi_p^{i+1}(T) - p\phi_p^i(T),$$

which shows that the family $\{\phi_p^i f_p^1\}_{i \geq 0}$ is a linearly independent collection of characters. Thus, $\tilde{f}$ is in the span of $\{\phi_p^i f_p^1 \circ\}_{i=0}^{n-1}$, and therefore, f must be in the $R[\phi_p]$-span of f_p^1. This finishes the proof. □

Now if $G = E$ is an elliptic curve over R, then $\mathcal{O}_{\delta_p}^\infty(E)$ is also a free $R[\phi_p]$-module of rank 1. But the basis is a character of order 1 or 2 depending on whether E has a canonical lift or not. An elliptic curve E over R is called a canonical lift if there exists a morphism of schemes $E \to E$ over $\mathbb{Z}_p$ whose reduction mod p is the "absolute Frobenius." We simply formulate the theorem for completeness. The reader is referred to [8] for its proof, whose discussion exceeds the goals we have for this monograph.

Theorem 6.24 (Buium [8]) *If E is any elliptic curve over R, then there exists a a nonzero character $f^2 : E(R) \to R$ of order two whose group of solutions contains $\cap p^n E(R)$ as a subgroup of finite index, and f^2 generates $\mathcal{O}_{\delta_p}^\infty(E)$ as an $R[\phi_p]$-module. If E has ordinary reduction* and is a canonical lift of its reduction, there exists a nonzero δ_p-character $f^1 : E(R) \to R$ of order 1, unique up to multiplication by a constant, and in this case, f^2 can be written as a linear combination of f^1 with coefficients taken out of $R[\phi_p]$, and the character f^1 generates the free $R[\phi_p]$-module $\mathcal{O}^\infty_{\delta_p}(E)$.

The character f^1_p in Theorem 6.23 is the arithmetic analogue of the Kolchin logarithmic derivative [30], while the character f^2 in Theorem 6.24 is the arithmetic analogue of the Manin map [37].

6.4 Multiple primes II

In extending the theory above to the partial differential setting, there are two ways of proceeding, one already explained in §5.1. These ways are obtained when we consider geometric directions of differentiation in addition to the arithmetic one, or when we consider the various arithmetic directions associated to a set of multiple primes. These two cases can be briefly illustrated using first the simple minded extension of the analogy between $(\mathbb{C}[x], \frac{d}{dx})$ and $(\mathbb{Z}, \delta_p)$ used earlier.

Indeed, we now consider the polynomial ring $\mathbb{C}[x_1, x_2]$. Given primes p_1, p_2, we may propose as the analogue of the algebra $(\mathbb{C}[x_1, x_2], \partial_{x_1}, \partial_{x_2})$, $\partial_{x_i} := \frac{\partial}{\partial x_i}$, $i = 1, 2$, the triple $(\mathbb{Z}, \delta_{p_1}, \delta_{p_2})$, where the two arithmetic derivatives $\delta_{p_1}, \delta_{p_2} : \mathbb{Z} \to \mathbb{Z}$ are defined by $\delta_{p_i} n := \dfrac{n - n^{p_i}}{p_i}$, $i = 1, 2$. This captures the essence of the case of several arithmetic directions. On the other hand, we could simply extend the theory to the partial differential case by taking as the analogue of $(\mathbb{C}[x_1, x_2], \partial_{x_1}, \partial_{x_2})$ the triple $(\mathbb{Z}[q], \delta_p, \delta_q)$, where $\mathbb{Z}[q]$ is a polynomial ring in the indeterminate q, and δ_p and δ_q are defined by $\delta_p(\sum a_n q^n) := \dfrac{(\sum a_n q^{pn}) - (\sum a_n q^n)^p}{p}$ and $\delta_q(\sum a_n q^n) := q\dfrac{d}{dq}\left(\sum a_n q^n\right) = \sum n a_n q^n$, respectively. In this context, we think of p and q as arithmetic and geometric directions, respectively. The latter case is beyond the scope of this monograph, and we refer the interested reader directly to the source [15, 16], where the needed jet spaces foundational to the theory are introduced, and complete statements on the characters on one dimensional group schemes are made, paralleling the ones above for a single prime. On the other hand, the first of the cases above is also outside the scope of this monograph, but we started in §5.1 to explain the difficulties in going beyond a single prime, and we shall elaborate on it further in this section, in a rather sketchy form. The interested reader may find the complete details in [17].

Given a family $\mathcal{P} = \{p_1, \ldots, p_d\}$ of primes, let A_0 be a $\delta_{\mathcal{P}}$-ring, for instance, the ring of Example 5.8. We denote by K_0 its fraction field. We let A be the ring $A = A_0[[q]]$ of power series in q, and define a family of homomorphism $\phi_{p_k} : A \to A$, $1 \leq k \leq d$, by

$$\phi_{p_k}\left(\sum c_n q^n\right) = \sum \phi_{p_k}(c_n) q^{np_k}.$$

Then A is a $\delta_{\mathcal{P}}$-ring also with respect to the p_k-derivations δ_{p_k} associated to the homomorphism ϕ_{p_k}s.

Given an n-tuple of variables x, we consider n-tuples of variables x_i indexed by vectors $i = (i_1, \ldots, i_d)$ in $\mathbb{Z}_{\geq 0}^d$ such that $x = x_{(0,\ldots,0)}$. We set $\delta_{\mathcal{P}}^i = \delta_{p_1}^{i_1} \ldots \delta_{p_d}^{i_d}$, and $\phi_{\mathcal{P}}^i = \phi_{p_1}^{i_1} \ldots \phi_{p_d}^{i_d}$.

Let $K_0\{x\}$ be the ring of polynomials

$$K_0\{x\} := K_0[x_i : i \geq 0]$$

with K_0-coefficients in the variables x_i with $i \in \mathbb{Z}_{\geq 0}$. The homomorphisms $\phi_{p_k} : A_0 \to A_0$ are extended to ring endomorphisms $\phi_{p_k} : K_0\{x\} \to K_0\{x\}$ by $\phi_{p_k}(x_i) = x_{i+e_k}$, so that we have that $x_i = \phi_{\mathcal{P}}^i x$ for all i. Clearly $\phi_{p_k}\phi_{p_l}(a) = \phi_{p_l}\phi_{p_k}(a)$ for all $a \in K_0\{x\}$, all k, l. If we consider the p_k-derivations $\delta_{p_k} : K_0\{x\} \to K_0\{x\}$ associated to the ϕ_{p_k}s, then $K_0\{x\}$ is a $\delta_{\mathcal{P}}$-ring that is generated as a K_0-algebra by the elements $\delta_{\mathcal{P}}^i x$, $i \geq 0$:

$$K_0\{x\} = K_0[\delta_{\mathcal{P}}^i x : i \geq 0].$$

We define the ring of $\delta_{\mathcal{P}}$-*polynomials* $A_0\{x\}$ to be the A_0-subalgebra of $K_0\{x\}$ generated by all the elements $\delta_{\mathcal{P}}^i x$:

$$A_0\{x\} := A_0[\delta_{\mathcal{P}}^i x : i \geq 0].$$

The ring $A_0\{x\}$ is strictly larger than the ring $A_0[x_i : i \geq 0]$. And the family $\{\delta_{\mathcal{P}}^i x : i \geq 0\}$ is algebraically independent over A_0, so $A_0\{x\}$ is a ring of polynomials in the variables $\delta_{\mathcal{P}}^i x$.

Lemma 6.25 *We have that* $\delta_{p_k} A_0\{x\} \subset A_0\{x\}$ *for* $k = 1, \ldots, d$, *and so* $A_0\{x\}$ *is in a natural manner a* $\delta_{\mathcal{P}}$-*ring.*

Proof. By the definition of a p-derivation (see (5.2)), the sets $S_k := \{a \in K_0\{x\} : \delta_{p_k} a \in A_0\{x\}\}$ are A_0-subalgebras of $K_0\{x\}$. Therefore, it suffices to show that $\delta_{\mathcal{P}}^i x \in S_k$ for all i and k. This can be done using the the commutation relations (5.9) via an induction argument on $(i, k) \in \mathbb{Z}_{\geq 0}^d \times \mathbb{Z}_{\geq 0}$ with respect to the lexicographic order.   □

Proceeding as before the statement of the lemma, we see that the system

$$A_0[\delta_{\mathcal{P}}^i x : i \leq r]$$

has a natural structure of $\delta_{\mathcal{P}}$-prolongation system.

Example 6.26   Let T be a tuple of indeterminates, and let

$$A^r = A_0[[\delta_{\mathcal{P}}^i T : i \leq r]] .$$

Then, the structure of $\delta_{\mathcal{P}}$-prolongation sequence above induces a structure of $\delta_{\mathcal{P}}$-prolongation sequence on the sequence of rings (A^r). For the p_k-derivation δ_{p_k} sends the ideal

$$I_r := (\delta_{\mathcal{P}}^i T : i \leq r) \subset A^r$$

into the ideal $I_{r+e_k} \subset A^{r+e_k}$.   □

We are now ready to define the $\delta_{\mathcal{P}}$-jet spaces. As in the case of a single prime [8] discussed above, we now have the following existence result for a universal prolongation sequence.

Proposition 6.27   *Let A^0 be a finitely generated A_0-algebra. Then there exists a $\delta_{\mathcal{P}}$-prolongation sequence A^* over A_0, with A^r finitely generated over A_0, satisfying the following property: for any $\delta_{\mathcal{P}}$-prolongation system B^* over A_0 and any A_0-algebra homomorphism $u : A^0 \to B^0$, there exists a unique morphism of $\delta_{\mathcal{P}}$-prolongation systems $u^* : A^* \to B^*$ such that $u^0 = u$.*

Proof. We express the finitely generated algebra A^0 as

$$A^0 = \frac{A_0[x]}{(f)}$$

for a tuple of indeterminates x, and a tuple of polynomials f. Then we set

$$A^r = \frac{A_0[\delta_{\mathcal{P}}^i x : i \leq r]}{(\delta_{\mathcal{P}}^i f : i \leq r)} .$$

We check easily that $A^* = (A^r)$ has the universality property in the statement.   □

Definition 6.28 Let X be an affine scheme of finite type over A_0. Let $A^0 = \mathcal{O}(X)$ and let A^r be as in Proposition 6.27. Then the scheme

$$J_{\mathcal{P}}^r(X) := \operatorname{Spec} A^r$$

is called the $\delta_{\mathcal{P}}$-jet space of order r of X. $\qquad\square$

By the universality property in Proposition 6.27, up to isomorphism the scheme $J_{\mathcal{P}}^r(X)$ depends on X alone, and is functorial in X: for any morphism $X \to Y$ of affine schemes of finite type, there are induced morphisms of schemes

$$J_{\mathcal{P}}^r(Y) \to J_{\mathcal{P}}^r(X).$$

Notice that when $\mathcal{P}$ consists of a single prime p, the p-adic completions of the schemes $J_p^r(X)$ are those introduced and studied in [8, 11, 6], and discussed earlier. For arbitrary $\mathcal{P}$, the schemes $J_{\mathcal{P}}^r(X)$ above were independently introduced by Borger in [4], where they are denoted by $W_{r*}(X)$.

Lemma 6.29 Let X be an affine scheme of finite type over A_0 and let $Y \subset X$ be a principal open set of X, $\mathcal{O}(Y) = \mathcal{O}(X)_f$. Then $\mathcal{O}(J_{\mathcal{P}}^r(Y)) \simeq \mathcal{O}(J_{\mathcal{P}}^r(X))_{f_r}$ where $f_r = \prod_{i \leq r} \phi_{\mathcal{P}}^i(f)$. In particular, the induced morphism $J_{\mathcal{P}}^r(Y) \to J_{\mathcal{P}}^r(X)$ is an open immersion whose image is principal, and if we view this morphism as an inclusion, and $Z \subset X$ is another principal open set, then we have that

$$J_{\mathcal{P}}^r(Y \cap Z) = J_{\mathcal{P}}^r(Y) \cap J_{\mathcal{P}}^r(Z).$$

Proof. We can check that $\mathcal{O}(J_{\mathcal{P}}^r(X))_{f_r}$ has the universality property of $\mathcal{O}(J_{\mathcal{P}}^r(Y))$. The δ_{p_k}-derivations on $\mathcal{O}(J_{\mathcal{P}}^r(X))_{f_r}$ are defined via the formula

$$\delta_{p_k}\left(\frac{a}{b}\right) = \frac{b^{p_k}\delta_{p_k}a - a^{p_k}\delta_{p_k}b}{b^p \phi_{p_k}(b)}.$$

$\qquad\square$

Definition 6.30 Let X be a scheme of finite type over A_0. An affine open covering

$$X = \bigcup_{i=1}^{m} X_i \qquad (6.6)$$

is called *principal* if

$$X_i \cap X_j \text{ is principal in both, } X_i \text{ and } X_j, \qquad (6.7)$$

for all $i, j = 1, \ldots, m$. If $c = \{X_i\}_{i=1}^m$ is a principal covering of X, we define the $\delta_{\mathcal{P}}$-*jet space of order* r *of* X *with respect to this covering* by gluing $J_{\mathcal{P}}^r(X_i)$ along $J_{\mathcal{P}}^r(X_i \cap X_j)$, and it is denoted by $J_{c,\mathcal{P}}^r(X)$.   □

If X is affine, we may use the *tautological covering* c of X, that is to say, the covering of X with a single open set, the set X itself. Then the space $J_{c,\mathcal{P}}^r(X)$ coincides with the space $J_{\mathcal{P}}^r(X)$ of Definition 6.28. Notice that any quasi-projective scheme X admits a principal covering.

In general, the schemes $J_{c,\mathcal{P}}^r(X)$ in the definition above depend on the covering c in a nontrivial manner. As shown and indicated already in [17], even though they have this deficiency, they suffice for the extension of the theory we seek. Because it is the case that the rings of global functions $\mathcal{O}(J_{c,\mathcal{P}}^r(X)^{\widehat{p_j}})$ on the p_j-adic completion $J_{c,\mathcal{P}}^r(X)^{\widehat{p_j}}$ of $J_{c,\mathcal{P}}^r(X)$ do not depend on the covering c, and are functorial in X. In fact, we have the following [17, Proposition 2.22].

Proposition 6.31   *Let us assume that* X *is a smooth scheme of finite type, quasi-projective and with connected geometric fibers over* $A_0 = \mathbb{Z}_{(\mathcal{P})} := \cap_{l=1}^d \mathbb{Z}_{(p_l)}$. *Let* c *and* c' *be two principal coverings of* X, *and* $J_{c,\mathcal{P}}^r(X)$ *and* $J_{c',\mathcal{P}}^r(X)$ *be the corresponding jet spaces. Then, there is a natural isomorphism*

$$\mathcal{O}(J_{c,\mathcal{P}}^r(X)^{\widehat{p_k}}) \simeq \mathcal{O}(J_{c',\mathcal{P}}^r(X)^{\widehat{p_k}}) .$$

We therefore drop the covering from the notation, and denote the isomorphism class of rings $\mathcal{O}(J_{c,\mathcal{P}}^r(X)^{\widehat{p_k}})$ simply by $\mathcal{O}(J_{\mathcal{P}}^r(X)^{\widehat{p_k}})$.

When $\mathcal{P}$ consists of a single prime p, we have interesting formal functions $f \in \mathcal{O}(J_p^r(X)^{\widehat{p}})$ for a handful of interesting situations. In the case of several primes, we would like to "glue together" family of these elements $f_{p_j} \in \mathcal{O}(J_{\mathcal{P}}^r(X)^{\widehat{p_j}})$, $j = 1, \ldots, d$. This gluing cannot be done directly since, for instance, in the case where X is affine, each f_{p_j} is a function on the "tubular neighborhood" $\mathrm{Spf}\, \mathcal{O}(J_{\mathcal{P}}^r(X))^{\widehat{p_j}}$ of $\mathrm{Spec}\, \mathcal{O}(J_{\mathcal{P}}^r(X)) \otimes \mathbb{F}_{p_j}$ in $\mathrm{Spec}\, \mathcal{O}(J_{\mathcal{P}}^r(X))$, and these tubular neighborhoods are disjoint. We solved this in [17] by defining the notion of the "analytic continuation" of the various elements on the family.

Definition 6.32   Let A be a ring, I be an ideal in A, and $\mathcal{P} = \{p_1, \ldots, p_d\}$ be a finite set of primes in $\mathbb{Z}$ that are noninvertible in A/I. We say that a family

$$f = (f_k) \in \prod_{k=1}^d A^{\widehat{p_k}} \tag{6.8}$$

can be *analytically continued along* I if there exists $f_0 \in A^I$ such that the images of f_0 and f_k in the ring $A^{\widehat{(p_k, I)}}$ coincide for each $k = 1, \ldots, d$. In that case, we say that f_0 *represents* f. If $X = \operatorname{Spec} A$, we denote by $\mathcal{O}_{I, \mathcal{P}}(X)$ the ring of families (6.8) that can be analytically continued along I. $\square$

From this point on in the section, we take the $\delta_{\mathcal{P}}$-ring A_0 to be just $\mathbb{Z}_{(\mathcal{P})}$. Given a $\mathbb{Z}_{(\mathcal{P})}$-point $P : \operatorname{Spec} \mathbb{Z}_{(\mathcal{P})} \to X$, by the universality property we obtain a unique lift to a point $P^r : \operatorname{Spec} \mathbb{Z}_{(\mathcal{P})} \to J^r_{\mathcal{P}}(X)$ that is compatible with the action of $\delta_{\mathcal{P}}$.

Definition 6.33 Let X be an affine scheme of finite type over $\mathbb{Z}_{(\mathcal{P})}$, P be a $\mathbb{Z}_{(\mathcal{P})}$-point $P : \operatorname{Spec} \mathbb{Z}_{(\mathcal{P})} \to X$, and $P^r : \operatorname{Spec} \mathbb{Z}_{(\mathcal{P})} \to J^r_{\mathcal{P}}(X)$ be its unique lift compatible with the action of $\delta_{\mathcal{P}}$. We denote by $P^r \subset \mathcal{O}(J^r_{\mathcal{P}}(X))$ also the ideal of the image of P^r. By a $\delta_{\mathcal{P}}$-*function on X of order r that is analytically continued along P* we mean a family

$$f = (f_k) \in \prod_{k=1}^{d} \mathcal{O}(J^r_{\mathcal{P}}(X))^{\widehat{p_k}} \tag{6.9}$$

that can be analytically continued along P^r. We denote by $\mathcal{O}^r_{P, \mathcal{P}}(X)$ the ring of all $\delta_{\mathcal{P}}$-functions on X of order r that are analytically continued along P.

Notice then that $f = (f_k) \in \prod_{k=1}^{d} \mathcal{O}(J^r_{\mathcal{P}}(X))^{\widehat{p_k}}$ can be analytically continued along P^r if, and only if, there exists an element

$$f_0 \in \mathcal{O}(J^r_{\mathcal{P}}(X))^{\widehat{P^r}}$$

that represents f, that is to say, such that the images of f_0 and f_k in

$$\mathcal{O}(J^r_{\mathcal{P}}(X))^{\widehat{(p_k, P^r)}}$$

coincide for each $k = 1, \ldots, d$. Thus, the the ring of all $\delta_{\mathcal{P}}$-functions on X that are analytically continued along P is

$$\mathcal{O}^r_{P, \mathcal{P}}(X) := \mathcal{O}_{P^r, \mathcal{P}}(J^r_{\mathcal{P}}(X)).$$

We now have the following definition, which unravels the analytic continuation concept in a pragmatic way under some general hypotheses on X and P.

Definition 6.34 Let X be a smooth affine scheme over $\mathbb{Z}_{(\mathcal{P})}$. A $\mathbb{Z}_{(\mathcal{P})}$-point $P : \operatorname{Spec} \mathbb{Z}_{(\mathcal{P})} \to X$ of X is called *uniform* if there exists an étale map of $\mathbb{Z}_{(\mathcal{P})}$-algebras $\mathbb{Z}_{(\mathcal{P})}[T] \to \mathcal{O}(X)$, where T is a tuple of indeterminates, such that the ideal of the image of P in $\mathcal{O}(X)$ is generated by T. We refer to T as *uniform coordinates*. Then T is a regular sequence in $\mathcal{O}(X)$, and the graded ring associated to the ideal (T) in $\mathcal{O}(X)$ is isomorphic to $\mathbb{Z}_{(\mathcal{P})}[T]$. We have that

$$\mathcal{O}(X)^{\widehat{T}} \simeq \mathbb{Z}_{(\mathcal{P})}[[T]] \,,$$

and similarly that

$$\mathcal{O}(X)^{\widehat{(p_k,T)}} \simeq \mathbb{Z}_{p_k}[[T]] \,.$$

For a general scheme X, we say that a $\mathbb{Z}_{(\mathcal{P})}$-point P of X is *uniform* if there exists an affine open set $X' \subset X$ that contains P such that P is uniform in X'. $\square$

We now have the following.

Definition 6.35 Let X be a smooth quasi-projective scheme over $\mathbb{Z}_{(\mathcal{P})}$ with geometrically connected fibers, and let P be a uniform point in some affine open set X'. Let us denote by $\mathcal{O}^r_{P,\mathcal{P}}(X)$ the pre-image of $\mathcal{O}^r_{P,\mathcal{P}}(X')$ via the restriction map

$$\prod_{k=1}^{d} \mathcal{O}(J^r_{\mathcal{P}}(X)^{\widehat{p_k}}) \to \prod_{k=1}^{d} \mathcal{O}(J^r_{\mathcal{P}}(X')^{\widehat{p_k}}) = \prod_{k=1}^{d} \mathcal{O}(J^r_{\mathcal{P}}(X'))^{\widehat{p_k}} \,.$$

Elements of the ring $\mathcal{O}^r_{P,\mathcal{P}}(X)$ are referred to as $\delta_{\mathcal{P}}$-*functions of order r on X that are analytically continued along P*. We define the ring of $\delta_{\mathcal{P}}$-*functions on X that are analytically continued along P* by

$$\mathcal{O}^{\infty}_{P,\mathcal{P}}(X) := \varinjlim \mathcal{O}^r_{P,\mathcal{P}}(X) \,,$$

with its natural $\delta_{\mathcal{P}}$-ring structure. $\square$

Example 6.36 Let $X = \mathbb{A}^n = \operatorname{Spec} \mathbb{Z}_{(\mathcal{P})}[T]$ be the n-dimensional affine space over $\mathbb{Z}_{(\mathcal{P})}$, where T is an n-tuple of indeterminates. We let P be the zero section, and let t be the tuple of indeterminates $(\delta^i_{\mathcal{P}}T)_{i \leq r}$. If ord_{p_k} denotes the p_k-adic order, then

$$\mathcal{O}^r_{P,\mathcal{P}}(\mathbb{A}^n) \simeq \{ \textstyle\sum a_j t^j \in \mathbb{Z}_{(\mathcal{P})}[[t]] \ : \ \lim_{|j| \to \infty} \operatorname{ord}_{p_k} a_j \to \infty \text{ for each } k \} \,.$$

Since in this case $\operatorname{Spec} \mathbb{Z}_{(\mathcal{P})} \subset X$, the lift to $J^r_{\mathcal{P}}(X)$ of other $\mathbb{Z}_{(\mathcal{P})}$-points, and the description of the $\delta_{\mathcal{P}}$ functions that can be analytically continued along them, is carried in a similar manner via a "translation" of the zero section. $\square$

We end our discussion of the theory for several primes by stating the main results about group characters and the analogues of those results of Buium given in §6.3 for the case of a single prime. Again, this is just a sketch, and the interested reader is referred to [17] for complete results.

Let G be a smooth group scheme over $\mathbb{Z}_{(\mathcal{P})}$ with multiplication μ : $G \times G \to G$, where, $\times = \times_{\mathbb{Z}_{(\mathcal{P})}}$. If G is affine, by the universality property we have that $J_{\mathcal{P}}^r(G)$ is a group scheme over $\mathbb{Z}_{(\mathcal{P})}$. In the nonaffine case, the covering dependent definitions may create problems, but we attach to this general situation a formal group law as follows.

Let Z be the identity $\operatorname{Spec} \mathbb{Z}_{(\mathcal{P})} \to G$, and assume Z is uniform, with uniform coordinates T (see Definition 6.34). This condition on Z is trivially satisfied in the cases where $G = \mathbb{G}_a$, $G = \mathbb{G}_m$, and $G = E$, an elliptic curve. For arbitrary G, the condition that Z be uniform is not restrictive whenever we choose the primes in $\mathcal{P}$ to be sufficiently large.

Under the conditions above, $Z \times Z \subset G \times G$ is a uniform point with uniform coordinates T_1, T_2 induced by T. We have an induced homomorphism $\mathbb{Z}_{(\mathcal{P})}[[T]] \to \mathbb{Z}_{(\mathcal{P})}[[T_1, T_2]]$ that sends the ideal (T) into the ideal (T_1, T_2). By the universality property applied to the restriction $\mathbb{Z}_{(\mathcal{P})}[T] \to \mathbb{Z}_{(\mathcal{P})}[[T_1, T_2]]$, we obtain morphisms

$$\mathbb{Z}_{(\mathcal{P})}[\delta_{\mathcal{P}}^i T : i \leq r] \to \mathbb{Z}_{(\mathcal{P})}[[\delta_{\mathcal{P}}^i T_1, \delta_{\mathcal{P}}^i T_2 : i \leq r]]$$

that send the ideal generated by the variables into the corresponding ideal generated by the variables. Thus, we have an induced morphism

$$\mathbb{Z}_{(\mathcal{P})}[[\delta_{\mathcal{P}}^i T : i \leq r]] \to \mathbb{Z}_{(\mathcal{P})}[[\delta_{\mathcal{P}}^i T_1, \delta_{\mathcal{P}}^i T_2 : i \leq r]].$$

We call $\mathcal{G}^r$ the image of the variables $\{\delta_{\mathcal{P}}^i T : i \leq r\}$ under this last homomorphism. Then the tuple $\mathcal{G}^r$ is a formal group law over $\mathbb{Z}_{(\mathcal{P})}$.

Definition 6.37 Let G be a quasi-projective smooth group scheme over $\mathbb{Z}_{(\mathcal{P})}$ with geometrically connected fibers. Let us assume that the identity is a uniform point. Then there are homomorphisms

$$\mu^*, pr_1^*, pr_2^* : \mathcal{O}_{Z,\mathcal{P}}^r(G) \to \mathcal{O}_{Z \times Z, \mathcal{P}}^r(G \times G)$$

induced by the product μ and the two projections. We say that a $\delta_{\mathcal{P}}$-function $f \in \mathcal{O}_{Z,\mathcal{P}}^r(G)$ of order r on G that is analytically continued along Z is a $\delta_{\mathcal{P}}$-*character of order r on G* if

$$\mu^* f = pr_1^* f + pr_2^* f.$$

We denote by $\mathcal{X}_{\mathcal{P}}^r(G)$ the group of $\delta_{\mathcal{P}}$-characters of order r on G. We define the *group of $\delta_{\mathcal{P}}$-characters* on G to be

$$\mathcal{X}_{\mathcal{P}}^{\infty}(G) := \varinjlim \mathcal{X}_{\mathcal{P}}^{r}(G) \,.$$

$\square$

The condition that f be a $\delta_{\mathcal{P}}$-character of order r on G is that there exists

$$f_0 \in \mathbb{Z}_{(\mathcal{P})}[[\delta_{\mathcal{P}}^i T : i \leq r]]$$

that represents f such that

$$f_0(\mathcal{G}^r(T_1, T_2)) = f_0(T_1) + f_0(T_2) \,. \tag{6.10}$$

Here, $\mathcal{G}^r$ is the corresponding formal group law, and $f_0(T)$ stands for $f_0(\ldots, \delta_{\mathcal{P}}^i T, \ldots)$.

The group $\mathcal{X}_{\mathcal{P}}^{r}(G)$ of $\delta_{\mathcal{P}}$-characters of order r on G is a subgroup of the additive group of the ring $\mathcal{O}_{Z,\mathcal{P}}^{r}(G)$. The group $\mathcal{X}_{\mathcal{P}}^{\infty}(G)$ of $\delta_{\mathcal{P}}$-characters on G is a subgroup of $\mathcal{O}_{Z,G}^{\infty}(G)$.

For any $\delta_{\mathcal{P}}$-character $\psi \in \mathcal{X}_{\mathcal{P}}^{r}(G)$ and any $\delta_{\mathcal{P}}$-ring A, there is an induced group homomorphism

$$\psi_A : G(A_{\mathcal{P}}) \to A_{\mathcal{P}} \,, \tag{6.11}$$

where $A_{\mathcal{P}}$ is viewed as a group with respect to addition. Thus, we may speak of the *group of solutions* $\mathrm{Ker}\,\psi_A$ of the character ψ. The mapping $A \mapsto \psi_A$ is functorial in A.

Definition 6.38 Let P be a $\mathbb{Z}_{(\mathcal{P})}$-point of G. We say that a $\delta_{\mathcal{P}}$-character $\psi \in \mathcal{X}_{\mathcal{P}}^{r}(G)$ can be *analytically continued along P* if $\psi \in \mathcal{O}_{P,\mathcal{P}}^{r}(G)$. $\square$

We now exhibit the $\delta_{\mathcal{P}}$-characters for the groups $\mathbb{G}_a$ and $\mathbb{G}_m$, respectively, paralleling the discussion in §6.3 for a single prime:

The additive group. Let us consider the additive group scheme over $\mathbb{Z}_{(\mathcal{P})}$,

$$\mathbb{G}_a := \mathrm{Spec}\,\mathbb{Z}_{(\mathcal{P})}[x] \,.$$

The zero section is uniform, with uniform coordinate $T = x$. We have that

$$\mathcal{O}(J_{\mathcal{P}}^{r}(\mathbb{G}_a)) = \mathbb{Z}_{(\mathcal{P})}[\delta_{\mathcal{P}}^i x : i \leq r] \,.$$

Let us consider the polynomial ring

$$\mathbb{Z}_{(\mathcal{P})}[\phi_{\mathcal{P}}] := \mathbb{Z}_{(\mathcal{P})}[\phi_{p_l} : l \in I] = \sum_{n \in \mathcal{N}} \mathbb{Z}_{(\mathcal{P})}\phi_n \,,$$

where $I := \{1, \ldots, d\}$, $\mathcal{N}$ is the monoid of the natural numbers generated by $\mathcal{P}$, and the ϕ_{p_l}s are commuting variables. If $i = (i_1, \ldots, i_d) \in \mathbb{Z}_{\geq 0}^d$, we set $\mathcal{P}^i = p_1^{i_1} \ldots p_d^{i_d}$. If $n = \mathcal{P}^i$ we set $\phi_n = \phi_{\mathcal{P}}^i = \phi_{p_1}^{i_1} \ldots \phi_{p_d}^{i_d}$. The ring $\mathbb{Z}_{(\mathcal{P})}[\phi_{\mathcal{P}}]$ is then viewed as the *ring of symbols* of the arithmetic partial differential equations that we define. For $r \in \mathbb{Z}_{\geq 0}^d$, we consider the symbol

$$\bar{\psi} := \sum_{n | \mathcal{P}^r} c_n \phi_n \in \mathbb{Z}_{(\mathcal{P})}[\phi_{\mathcal{P}}] \,.$$

We may consider the element $\psi = \bar{\psi} x \in \mathcal{O}(J^r(\mathbb{G}_a))$, and using the diagonal embedding, identify it with an element

$$\psi \in \prod_{k=1}^d \mathcal{O}(J_{\mathcal{P}}^r(\mathbb{G}_a))^{\widehat{p_k}} \,.$$

This $\bar{\psi}$ clearly defines a $\delta_{\mathcal{P}}$-character on $\mathbb{G}_a$. In what follows, ψ and $\bar{\psi}$ shall be identified with each other.

The following result for the additive group is simple. Later on, a less elementary analogue for the multiplicative group $\mathbb{G}_m$ will follow. We will not discuss the multiple prime situation for elliptic curves here.

Theorem 6.39 *Let ψ be a $\delta_{\mathcal{P}}$-character on $\mathbb{G}_a$ of order r. Then*

1. ψ can be uniquely written as $\psi = \bar{\psi} x$ with

$$\bar{\psi} = \left(\sum_n c_n \phi_n \right) x \,, \quad c_n \in \mathbb{Z}_{(\mathcal{P})} \,.$$

2. ψ can be analytically continued along any $\mathbb{Z}_{(\mathcal{P})}$-point P of $\mathbb{G}_a$.

Proof. Let $\psi \in \mathcal{X}_{\mathcal{P}}^r(\mathbb{G}_a)$, and suppose ψ is represented by a series $\psi_0 \in \mathbb{Z}_{(\mathcal{P})}[[\delta_{\mathcal{P}}^i x : i \leq r]]$. We view this representative ψ_0 as an element of

$$\mathbb{Q}[[\delta_{\mathcal{P}}^i x : i \leq r]] = \mathbb{Q}[[\phi_{\mathcal{P}}^i x : i \leq r]] \,.$$

Notice that we have that

$$\begin{aligned} \psi_0(\ldots, \phi_{\mathcal{P}}^i x_1 + \phi_{\mathcal{P}}^i x_2, \ldots) &= \psi_0(\ldots, \phi_{\mathcal{P}}^i (x_1 + x_2), \ldots) \\ &= \psi_0(\ldots, \phi_{\mathcal{P}}^i x_1, \ldots) + \psi_0(\ldots, \phi_{\mathcal{P}}^i x_2, \ldots) \,. \end{aligned}$$

Then $\psi_0 = \sum c_n \phi_n x$, where $c_n \in \mathbb{Q}$.

We have that

$$\phi_n x \equiv x^n \bmod (\delta_{\mathcal{P}}^i x : i \leq r)$$

in the ring $\mathbb{Q}[\delta_{\mathcal{P}}^i x : i \leq r]$. Therefore, we obtain that

$$(\psi_0)_{|\delta_{\mathcal{P}}^i x = 0;\, 0 \neq i \leq r} = \sum c_n x^n \,.$$

It follows that $c_n \in \mathbb{Z}_{(\mathcal{P})}$, which completes the proof of the first assertion.

The second assertion is clear. We merely refer to Example 6.36 for (some) details. $\qquad\square$

Corollary *The group of $\delta_{\mathcal{P}}$-characters $\mathfrak{X}_{\mathcal{P}}^\infty(\mathbb{G}_a)$ is a free $\mathbb{Z}_{(\mathcal{P})}[\phi_{\mathcal{P}}]$-module of rank one with basis x.*

The multiplicative group Let us now consider the multiplicative group scheme over $\mathbb{Z}_{(\mathcal{P})}$,

$$\mathbb{G}_m := \operatorname{Spec} \mathbb{Z}_{(\mathcal{P})}[x, x^{-1}] \,.$$

The zero section is uniform, with uniform coordinate $T = x - 1$.

The formal group law $\mathcal{G}^0$ corresponding to T is

$$\mathcal{G}^0(T_1, T_2) = T_1 + T_2 + T_1 T_2 \,,$$

and the logarithm of this formal group law is given by the series

$$l_{\mathbb{G}_m}(T) = \sum_{n=1}^{\infty} (-1)^{n-1} \frac{T^n}{n} \in \mathbb{Q}[[T]] \,.$$

By Lemma 6.29, we have that

$$\mathcal{O}(J_{\mathcal{P}}^r(\mathbb{G}_m)) = \mathbb{Z}_{(\mathcal{P})}\left[\delta_{\mathcal{P}}^i x, \frac{1}{\phi_{\mathcal{P}}^i(x)} : i \leq r\right] \,.$$

For each k, we consider the series

$$\psi_{p_k} = \psi_{p_k}^1 := \sum_{n=1}^{\infty} (-1)^{n-1} \frac{p_k^{n-1}}{n} \left(\frac{\delta_{p_k} x}{x^{p_k}}\right)^n \in \mathcal{O}(J_{\mathcal{P}}^{e_k}(\mathbb{G}_m))^{\widehat{p_k}} \,,$$

and the element

$$\bar{\psi}_{\bar{p}_k} := \prod_{l \in I_k} \left(1 - \frac{\phi_{p_l}}{p_l}\right) = \sum_{n \in \mathcal{N}_k} \frac{\mu(n)}{n} \phi_n \in \mathbb{Z}_{(p_k)}[\phi_{p_l} : l \in I_k] \,.$$

In this expression, we have set $I_k := \{1, \ldots, d\} \backslash \{k\}$, $\mathcal{N}_k$ is the monoid of the natural numbers generated by $\bar{p}_k := \mathcal{P} \backslash \{p_k\}$, μ is the Möbius function and the ϕ_{p_l}s are a set of d-variables that commute among themselves. Let us then consider the family

$$(\bar{\psi}_{\bar{p}_k} \psi_{p_k}) \in \prod_{k=1}^{d} \mathcal{O}(J_{\mathcal{P}}^e(\mathbb{G}_m))^{\widehat{p_k}} \,,$$

where $e = e_1 + \cdots + e_d = (1, \ldots, 1)$. As we see right away, this family is a $\delta_{\mathcal{P}}$-character that we shall denote by

$$\psi_m^e \in \mathfrak{X}_{\mathcal{P}}^e(\mathbb{G}_m). \tag{6.12}$$

Theorem 6.40 *The family $(\bar{\psi}_{\bar{p}_k}\psi_{p_k})$ is a $\delta_{\mathcal{P}}$-character of order e on $\mathbb{G}_m$.*

Proof. We begin by observing that

$$\delta_{p_k}(1 + T) = \delta_{p_k}T + c_{p_k}(1, T) \in (T, \delta_{p_k}T) \subset \mathbb{Z}_{p_k}[[T, \delta_{p_k}T]], \tag{6.13}$$

where $c_p(x, y)$ is the polynomial (5.1) associated with p. Then the image of $\bar{\psi}_{\bar{p}_k}\psi_{p_k}$ in $\mathbb{Q}_{p_k}[[\delta_{\mathcal{P}}^i T : i \leq e]]$ is equal the following series:

$$
\begin{aligned}
\bar{\psi}_{\bar{p}_k}\left(\sum_{n=1}^{\infty} \frac{(-p_k)^{n-1}}{n}\left(\frac{\delta_{p_k}(1+T)}{(1+T)^{p_k}}\right)^n\right) &= \frac{1}{p_k}\bar{\psi}_{\bar{p}_k}\left(\sum_{n=1}^{\infty} \frac{(-1)^{n-1}}{n}\left(\frac{\phi_{p_k}(1+T)}{(1+T)^{p_k}} - 1\right)^n\right) \\
&= \frac{1}{p_k}\bar{\psi}_{\bar{p}_k}l_{\mathbb{G}_m}\left(\frac{\phi_{p_k}(1+T)}{(1+T)^{p_k}} - 1\right) \\
&= \frac{1}{p_k}\bar{\psi}_{\bar{p}_k}(\phi_{p_k} - p_k)l_{\mathbb{G}_m}(T) \\
&= -\left(\prod_{l=1}^{d}(1 - \frac{\phi_{p_l}}{p_l})\right)l_{\mathbb{G}_m}(T).
\end{aligned}
$$

Notice that by (6.13), this series is convergent in the topology given by the maximal ideal of the ring. We shall denote this series by ψ_0^e.

We have that ψ_0^e has coefficients in $\mathbb{Q} \cap \mathbb{Z}_{p_k} = \mathbb{Z}_{(p_k)}$, and is the same for all $k = 1, \ldots, d$. Hence, ψ_0^e has coefficients in $\mathbb{Z}_{(\mathcal{P})}$, and it represents the family $(\bar{\psi}_{\bar{p}_k}\psi_{p_k})$. We also have that ψ_0^e satisfies the condition (6.10) because

$$
\begin{aligned}
\psi_0^e(\mathcal{G}^e(T_1, T_2)) &= -\left(\prod_{l=1}^{d}(1 - \frac{\phi_{p_l}}{p_l})\right)\left(l_{\mathbb{G}_m}(T)(\mathcal{G}^0(T_1, T_2))\right) \\
&= -\left(\prod_{l=1}^{d}(1 - \frac{\phi_{p_l}}{p_l})\right)\left(l_{\mathbb{G}_m}(T_1) + l_{\mathbb{G}_m}(T_2)\right) \\
&= \psi_0^e(T_1) + \psi_0^e(T_2).
\end{aligned}
$$

This completes the proof. $\square$

Remark 6.41 Let us consider two primes, so $d = 2$. Then we have that the character $\psi_m^{(1,1)}$ above has the factorization

$$-\left(1 - \frac{\phi_{p_1}}{p_1}\right)\left(1 - \frac{\phi_{p_2}}{p_2}\right)\sum(-1)^{n-1}\frac{T^n}{n} \in \mathbb{Q}[[T, \delta_{p_1}T, \delta_{p_2}T, \delta_{p_1}\delta_{p_2}T]].$$

This is very much analogous to the factorization of the 2-dimensional positive Laplacian when it acts on the logarithm of a function. Indeed, we have

$$\frac{1}{4}\Delta \log u = -\partial_z \partial_{\bar{z}} \log u \,,$$

which can be written as

$$-\partial_z \left(\frac{\partial_{\bar{z}} u}{u} \right) = -\partial_{\bar{z}} \left(\frac{\partial_z u}{u} \right) \,,$$

Here, $z = x + iy$ is the complex coordinate on $\mathbb{C} = \mathbb{R}^2$, and $\Delta = -(\partial_x^2 + \partial_y^2)$.

The analogy above is of limited scope. For it is not clear how to tie up different primes through a conjugation operation. However, there is another way of looking at the factorization above for $\psi_m^{(1,1)}$, and in this other way, we can consider quite naturally the character ψ_m^e for an arbitrary number of primes. For the product $-\prod_{j=1}^d \left(1 - \frac{\phi_{p_j}}{p_j}\right)$ can be thought of as the factorization of a d-th order operator whose symbol decomposes into the product of linear factors with distinct roots along the various arithmetic directions given by the primes. In the classical case, these are of course the strictly hyperbolic differential operators. In this sense, the arithmetic Laplacians of [17] exhibit a hyperbolic behaviour. The pursue of this idea might be worthwhile to undertake in the future.   □

We now show that $\psi_m^e \in \mathcal{X}_{\mathcal{P}}^e(\mathbb{G}_m)$ generates the space of all $\delta_{\mathcal{P}}$-characters of $\mathbb{G}_m$, and determine all of the $\delta_{\mathcal{P}}$-characters ψ that can be analytically continued along any given $\mathbb{Z}_{(\mathcal{P})}$-point P of $\mathbb{G}_m$.

Theorem 6.42   *Let ψ be a $\delta_{\mathcal{P}}$-character of order r on $\mathbb{G}_m$. Then*

1. ψ can be uniquely written as

$$\psi = \left(\sum_n c_n \phi_n\right)\psi_m^e, \quad c_n \in \mathbb{Z}_{(\mathcal{P})}\,.$$

2. ψ can be analytically continued along a $\mathbb{Z}_{(\mathcal{P})}$-point P of $\mathbb{G}_m$ if, and only if, either P is a torsion point or $\sum_n c_n = 0$.

Proof. Let ψ_0 be the series representing ψ. Then we have an equality of the form

$$\psi_0 = \left(\sum_{n|\mathcal{P}^r} d_n \phi_n\right) l(T)$$

in $\mathbb{Q}[[\delta_{\mathcal{P}}^i T : i \leq r]]$ for $d_n \in \mathbb{Q}$.

Indeed let $e(T) \in \mathbb{Q}[[T]]$ be the compositional inverse of $l(T)$. Then the series

$$\Theta(\ldots, \delta_{\mathcal{P}}^i(T), \ldots) := \psi_0(\ldots, \delta_{\mathcal{P}}^i(e(T)), \ldots)$$

satisfies the identity

$$\Theta(\ldots, \delta_{\mathcal{P}}^i(T_1 + T_2), \ldots) = \Theta(\ldots, \delta_{\mathcal{P}}^i T_1, \ldots) + \Theta(\ldots, \delta_{\mathcal{P}}^i T_2, \ldots).$$

By the argument used in the proof of Theorem 6.39, we conclude that $\Theta = \sum d_n \phi_n(T)$, with $d_n \in \mathbb{Q}$, which completes the proof of the assertion.

By this preliminary result, if we set $\delta_{\mathcal{P}}^i T = 0$ for $i \neq 0$, we obtain that

$$\left(\sum d_n \phi_n \right) \star l(T) \in \mathbb{Z}_{(\mathcal{P})}[[T]],$$

where $\phi_n \star T := T^n$.

Now we can see that if a polynomial $\Lambda = \sum l_n \phi_n \in \mathbb{Q}[\phi_{p_1}, \ldots, \phi_{p_s}]$ satisfies that

$$\Lambda \star l(T) \in \mathbb{Z}_{p_k}[[T]] \otimes \mathbb{Q}$$

for some $k \in \{1, \ldots, s\}$, then Λ is divisible in the ring $\mathbb{Q}[\phi_{p_1}, \ldots, \phi_{p_s}]$ by $\phi_{p_k} - p_k$.

Indeed, let us divide Λ by $\phi_{p_k} - p_k$ in $\mathbb{Q}[\phi_{p_1}, \ldots, \phi_{p_s}]$, so that

$$\Lambda = \left(\sum a_n \phi_n \right)(\phi_{p_k} - p_k) + \sum b_n \phi_n,$$

for $a_n, b_n \in \mathbb{Q}$, and $b_n = 0$ if $p_k | n$. We prove that the remainder term above is identically zero by showing that $b_n = 0$ for all n.

Since $(\phi_{p_k} - p_k) \star l(T) \in \mathbb{Z}_{p_k}[[T]]$, it follows that $\left(\sum b_n \phi_n \right) \star l(T) \in \mathbb{Z}_{p_k}[[T]] \otimes \mathbb{Q}$. We may assume $\left(\sum b_n \phi_n \right) \star l(T) \in \mathbb{Z}_{p_k}[[T]]$. We have that

$$\left(\sum b_n \phi_n \right) \star l(T) = \sum_n \sum_m (-1)^{m-1} b_n \frac{T^{nm}}{m}. \tag{6.14}$$

Let us fix integers $n', \nu \geq 1$. By looking at the coefficient of $T^{n' p_k^\nu}$ in (6.14), we obtain that

$$- \sum_{n | n'} (-1)^{\frac{n'}{n}} \frac{n b_n}{n' p_k^\nu} \in \mathbb{Z}_{p_k}$$

because the equality $nm = n' p_k^\nu$ with $n \not\equiv 0 \bmod p_k$ implies that $m = \mu p_k^\nu$ and that $\mu \in \mathbb{Z}$, $n\mu = n'$, so $n | n'$, and $m = \frac{n'}{n} p_k^\nu$. Since n is odd for $b_n \neq 0$, it follows that

$$\sum_{n | n'} n b_n \in p_k^\nu \mathbb{Z}_{p_k},$$

and since this is true for all ν, we obtain that

$$\sum_{n|n'} nb_n = 0 \, .$$

As this is true for all n', we may now use the Mobius' inversion formula to conclude that $b_n = 0$ for all n. And this completes the proof of the assertion.

Applying this result, it follows that

$$\sum d_n \phi_n = \left(\sum c_n \phi_n \right) \prod_{k=1}^{d} \left(1 - \frac{\phi_{p_k}}{p_k} \right)$$

for some $c_n \in \mathbb{Q}$. We may proceed by induction on n to show that $c_n \in \mathbb{Z}_{(\mathcal{P})}$ for all n. This completes the proof of the first part of the Theorem.

For the proof of the second part, let $\tau : \mathbb{G}_m \to \mathbb{G}_m$ be the translation defined by the inverse of P, and let τ^* be the automorphism defined by τ on the various rings of functions. Since $\psi \in \mathcal{O}^r_{Z,\mathcal{P}}(\mathbb{G}_m)$ (recall that Z is the identity $\mathrm{Spec}\,\mathbb{Z}_{(\mathcal{P})} \to \mathbb{G}_m$), we have that $\tau^*\psi \in \mathcal{O}^r_{P,\mathcal{P}}(\mathbb{G}_m)$. But if $\psi = (\psi_k)$ then $\tau^*\psi_k = \psi_k - \psi_k(P_k)$ for all k. Now, if P is torsion or if $\sum_n c_n = 0$, it is then clear that $\psi_k(P_k) = 0$. So $\tau^*\psi_k = \psi_k$, hence $\psi \in \mathcal{O}^r_{P,\mathcal{P}}(\mathbb{G}_m)$.

Conversely, let P be non-torsion and suppose that $a \in \mathbb{Z}^\times_{(\mathcal{P})}$ is a given number. Let $\sum_n c_n \neq 0$, and assume that $\psi \in \mathcal{O}^r_{P,\mathcal{P}}(\mathbb{G}_m)$. We derive a contradiction. For let $p = p_1$ and $b := a^{p-1} \in 1 + p\mathbb{Z}_{(p)}$, so $b \neq 1$. By Mahler's p-adic analogue of the Hermite-Lindemann theorem [35, 3]*, we have that $\log b \notin \mathbb{Q}$, where here $\log : 1 + p\mathbb{Z}_p \to p\mathbb{Z}_p$ is the p-adic logarithm. Since $\tau^*\psi \in \mathcal{O}^r_{P,\mathcal{P}}(\mathbb{G}_m)$ and $\psi \in \mathcal{O}^r_{P,\mathcal{P}}(\mathbb{G}_m)$, it follows that $\tau^*\psi - \psi \in \mathcal{O}^r_{P,\mathcal{P}}(\mathbb{G}_m)$. But $\tau^*\psi - \psi = (-\psi_k(a))$ so, in particular, $\psi_1(b) \in \mathbb{Q}$. But

$$\psi_1(b) = -\left(\sum_n c_n \right) \cdot \prod_{l=1}^{d} \left(1 - \frac{1}{p_l} \right) \cdot \log b \, .$$

Since $\sum_n c_n \neq 0$, it follows that $\log b \in \mathbb{Q}$, the desired contradiction. □

Consider the *augmentation ideal*

$$\mathbb{Z}_{(\mathcal{P})}[\phi_{\mathcal{P}}]^+ := \left\{ \sum_n c_n \phi_n \in \mathbb{Z}_{(\mathcal{P})}[\phi_{\mathcal{P}}] : \sum_n c_n = 0 \right\} \, .$$

---

* Roughly speaking this says that the p-adic exponential function is transcendental at nonzero algebraic arguments.

Corollary *The group of $\delta_{\mathcal{P}}$-characters $\mathcal{X}_{\mathcal{P}}^\infty(\mathbb{G}_m)$ is a free $\mathbb{Z}_{(\mathcal{P})}[\phi_{\mathcal{P}}]$-module of rank one with basis ψ_m^e. The group of $\delta_{\mathcal{P}}$-characters in $\mathcal{X}_{\mathcal{P}}^\infty(\mathbb{G}_m)$ that can be analytically continued along a given non-torsion point P of $\mathbb{G}_m$ is isomorphic with the augmentation ideal $\mathbb{Z}_{(\mathcal{P})}[\phi_{\mathcal{P}}]^+$ as a $\mathbb{Z}_{(\mathcal{P})}[\phi_{\mathcal{P}}]$-module. This group is the same for all non-torsion Ps.*

We now compute the group of solutions of the $\delta_{\mathcal{P}}$-character ψ_m^e in (6.12).

Theorem 6.43 *Let A be the $\delta_{\mathcal{P}}$-ring $\mathbb{Z}_{(\mathcal{P})}[\zeta_m]$ in Example 5.8, and let $\psi_{m,A}^e : \mathbb{G}_m(A_{\mathcal{P}}) = A_{\mathcal{P}}^\times \to A_{\mathcal{P}}$ be the homomorphism (6.11) induced by ψ_m^e. Then*

$$\operatorname{Ker} \psi_{m,A}^e = (A_{\mathcal{P}}^\times)_{tors}.$$

Here, Γ_{tors} denotes the torsion group of an Abelian group Γ.

Proof. The nontrivial inclusion to prove is that $\operatorname{Ker} \psi_{m,A}^e \subset (A_{\mathcal{P}}^\times)_{tors}$. Let us take $Q = (Q_k) \in \operatorname{Ker} \psi_{m,A}^e$ so that

$$\bar{\psi}_{\bar{p}_k}(\psi_{p_k}(Q_k)) = 0 \tag{6.15}$$

for all k. Here $Q_k \in A^{\widehat{p_k}} = A^{\widehat{P_{k1}}} \times A^{\widehat{P_{k2}}} \times \cdots$, where $p_k = P_{k1}P_{k2}\cdots$ is the prime decomposition of $p_k A$. In order to show that Q is torsion, we may replace Q by any of its powers. So we may assume that $Q_k \in 1 + p_k A^{\widehat{p_k}}$ for all k. Then (6.15) produces that

$$\left(\prod_{l=1}^d (\phi_{p_l} - p_l)\right) l_{\mathbb{G}_m}(Q_k - 1) = 0.$$

Now the map

$$A^{\widehat{p_k}} \to A^{\widehat{p_k}}$$
$$\beta \mapsto (\phi_{p_l} - p_l)\beta$$

is injective for all k, l, assertion that we prove below. Using this result, we conclude that $l_{\mathbb{G}_m}(Q_k - 1) = 0$, which implies that $Q_k = 1$ by the injectivity of $l_{\mathbb{G}_m} : p_k A^{\widehat{p_k}} \to p_k A^{\widehat{p_k}}$. This completes the proof of the Theorem.

For the proof of the assertion, let us assume that $(\phi_{p_l} - p_l)\beta = 0$. We also have that $\phi_{p_l}^M \beta = \beta$, $M := [\mathbb{Q}(\zeta_m) : \mathbb{Q}]$. Since the polynomials $\phi_{p_l} - p_l, \phi_{p_l}^M - 1 \in \mathbb{Q}[\phi_{p_l}]$ are coprime, it follows that $\beta = 0$, as desired. $\square$

Elliptic curves. We present the statement for elliptic curves for completeness.

Theorem 6.44   [17] *Let E be an elliptic curve over A with ordinary reduction at all the primes in $\mathcal{P}$. Then there exists a character ψ_E^{2e} of order $2e = (2, \ldots, 2)$, unique up to a unit in A, and such that every other $\delta_{\mathcal{P}}$-character on E is obtained in an appropriate sense from it.*

Remark 6.45   There is a factorization for ψ_E^{2e} analogous to the factorization for ψ_m^e discussed in Remark 6.41, but not as explicit. For we now have that

$$-\prod_{j=1}^{d} \left(1 - a_j \frac{\phi_{p_j}}{p_j} + p_j \left(\frac{\phi_{p_j}}{p_j} \right)^2 \right) l_E(T),$$

where $l_E(T)$ is the logarithm of the formal group law attached to E, and $a_j \in \mathbb{Z}$ is the trace of the p_j-power Frobenius on the reduction mod p_j of E. We now have that the character ψ_E^{2e} admits a factorization with factors that are operators of order 2 along each of the arithmetic directions given by the d primes in $\mathcal{P}$. This property makes of ψ_E^{2e} an operator that exhibits a hyperbolic nature, as in the case of the character ψ_m^e of Remark 6.41 that arose when treating the multiplicative group. In this sense, the order of ψ_E^{2e} should be taken to be the scalar $2d$, and not the multi-index $2e$.   □

7

Analyticity of arithmetic differential operators

We now retake the study of arithmetic differential operators over the ring $\mathbb{Z}_p$ by proving that any arithmetic differential operator

$$F(x, \delta x, \ldots, \delta^r x) : \mathbb{Z}_p \to \mathbb{Z}_p$$

is an analytic function in the sense of [36, 42]. The following Lemma lays the ground work to accomplish this task [14].

As a polynomial function in x, the arithmetic derivative $\delta(x)$ is a p-adically continuous function, and the same is true of general arithmetic differential operators, or δ-functions, on $\mathbb{Z}_p$. What we show here is that these functions hae another property, in addition to their mere continuity.

Lemma 7.1 *If $x = a + p^n u$ is an element of the disc $a + p^n\mathbb{Z}_p$, let us write $\delta^k x$ as a polynomial function in u of degree p^k,*

$$\delta^k x = \sum_{j=0}^{p^k} c_{a,j}^k u^j \,,$$

with coefficients $c_{a,j}^k \in \mathbb{Q}_p$. Then the following p-adic estimates hold:

1. $\left\| c_{a,0}^k \right\|_p \leq 1.$

2. $\left\| c_{a,1}^k \right\|_p = \dfrac{1}{p^{n-k}}.$

3. $\left\| c_{a,j}^k \right\|_p \leq \dfrac{1}{p^{(n-k+1)j-1}}, \quad 2 \leq j \leq p^k.$

Proof. The first of the estimates follows by the identity $c_{a,0}^k = \delta^k(a)$, which is a p-adic integer since δ ranges in $\mathbb{Z}_p$. We prove the remaining two estimates by induction on k.

103

The estimates are clear for $k = 0$. We assume now that they hold for $k - 1$. That is to say, we have

$$\delta^{k-1} x = \sum_{j=0}^{p^{k-1}} c_{a,j}^{k-1} u^j ,$$

where, for $j \geq 1$, the coefficients $c_{a,j}^{k-1}$ satisfy the estimates

$$\left\| c_{a,j}^{k-1} \right\|_p \leq \frac{1}{p^{(n-k+2)j-1}} ,$$

and the equality holds in the case when the index j is 1. By (5.5), we have that

$$\delta^k x = \frac{\sum_{j=0}^{p^{k-1}} c_{a,j}^{k-1} u^j - (\sum_{j=0}^{p^{k-1}} c_{a,j}^{k-1} u^j)^p}{p} .$$

We use the multinomial expansion in order to express the p-th power above as a polynomial in u. We obtain

$$\delta^k x = \sum_{j=0}^{p^{k-1}} \frac{c_{a,j}^{k-1}}{p} u^j - \sum \binom{p}{\alpha_0, \ldots, \alpha_{p^{k-1}}} \frac{(c_{a,0}^{k-1})^{\alpha_0} \cdots (c_{a,p^{k-1}}^{k-1})^{\alpha_{p^{k-1}}}}{p} u^{j_\alpha} ,$$

where for convenience we have set $j_\alpha = 0\alpha_0 + 1\alpha_1 + \cdots + p^{k-1}\alpha_{p^{k-1}}$, and the last sum is over all nonnegative multi-indices $\alpha = (\alpha_0, \ldots, \alpha_{p^{k-1}})$ of weight p. Notice that exponent j_α of u in that sum ranges in between 0, corresponding to the multi-index $\alpha = (p, 0, \ldots, 0)$, and p^k, corresponding to the multi-index $\alpha = (0, \ldots, 0, p)$. This shows that $\delta^k x$ is a polynomial of degree p^k in u, as desired. There remains only the estimation of the p-adic norm of its coefficients.

If we write

$$\delta^k x = \sum_{j=0}^{p^k} c_{a,j}^k u^j ,$$

then we have the relations

$$c_{a,j}^k = \frac{c_{a,j}^{k-1}}{p} - \sum_{j_\alpha = j} \binom{p}{\alpha_0, \ldots, \alpha_{p^{k-1}}} \frac{(c_{a,0}^{k-1})^{\alpha_0} \cdots (c_{a,p^{k-1}}^{k-1})^{\alpha_{p^{k-1}}}}{p}$$

for $0 \leq j \leq p^{k-1}$, or

$$c_{a,j}^k = - \sum_{j_\alpha = j} \binom{p}{\alpha_0, \ldots, \alpha_{p^{k-1}}} \frac{(c_{a,0}^{k-1})^{\alpha_0} \cdots (c_{a,p^{k-1}}^{k-1})^{\alpha_{p^{k-1}}}}{p}$$

for $p^{k-1} < j \leq p^k$. Since each multinomial coefficient is a rational integer, by the induction hypothesis applied to each of the coefficients $c_{a,l}^{k-1}$, we have that

$$\left\| \begin{pmatrix} p \\ \alpha_0, \ldots, \alpha_{p^{k-1}} \end{pmatrix} \frac{(c_{a,0}^{k-1})^{\alpha_0} \cdots (c_{a,p^{k-1}}^{k-1})^{\alpha_{p^{k-1}}}}{p} \right\|_p \leq \frac{p}{p^{(n-k+1)j_\alpha}} \, .$$

If $j_\alpha = 0\alpha_0 + 1\alpha_1 + \cdots + p^{k-1}\alpha_{p^{k-1}} = j$, the right hand side of this expression is simply $p^{-(n-k+1)j+1}$, and we may now estimate the p-adic norm of $c_{a,j}^k$ for any $j \geq 1$ using the expression for the coefficient $c_{a,j}^k$ given above. Indeed, by the non-Archimedean property of $\| \ \|_p$, and the estimate just derived, it follows that

$$\left\| c_{a,j}^k \right\|_p \leq \frac{1}{p^{(n-k+1)j-1}}$$

for any $j \geq 1$.

The estimate above is sharp for $j = 1$. Indeed, we have that

$$c_{a,1}^k = \frac{c_{a,1}^{k-1}}{p}(1 - p(c_{a,0}^{k-1})^{p-1}) \, , \tag{7.1}$$

and so

$$\left\| c_{a,1}^k \right\|_p = p \left\| c_{a,1}^{k-1} \right\|_p = \frac{p}{p^{n-k+1}} = \frac{1}{p^{n-k}} \, ,$$

as desired. This completes the proof. $\qquad\qquad\Box$

Remark 7.2 The coefficient $c_{a,1}^k$ is given by

$$c_{a,1}^k = p^{n-k} \prod_{l=0}^{k-1} (1 - p(\delta^l(a))^{p-1}) \, ,$$

as follows by the recursion formula (7.1). Since the $\delta^l(a)$s are all p-adic integers, by the non-Archimedean property of $\| \ \|_p$ applied to this expression, it follows easily that the p-adic norm of this coefficient is $1/p^{n-k}$. $\qquad\qquad\Box$

Theorem 7.3 *Any δ-function is an analytic function.*

Proof. Let $f(x) = F(x, \delta x, \ldots, \delta^r x)$ by an operator of order r given by the restricted power series $F \in \mathbb{Q}_p[[t_0, \ldots, t_r]]$. Thus,

$$f(x) = \sum_{\alpha=(\alpha_0, \ldots, \alpha_r)} a_\alpha x^{\alpha_0} (\delta x)^{\alpha_1} \cdots (\delta^r x)^{\alpha_r} \, ,$$

where $a_\alpha \to 0$ p-adically as $|\alpha| \to \infty$.

The family of discs $\{a+p^n\mathbb{Z}_p\}_{a=0}^{p^n-1}$ forms a covering of $\mathbb{Z}_p$. On each one of these discs, we consider the power series $f_a(u) = f(a+p^n u) \in \mathbb{Q}_p[[u]]$. We use Lemma 7.1 to show that this power series converges on $\mathbb{Z}_p$ for a suitable choice of n.

Indeed, we have that

$$f(a+p^n u) = \sum_{\alpha=(\alpha_0,\ldots,\alpha_r)} a_\alpha (\sum_{j_0=0}^{p^0} c_{a,j_0}^0 u^{j_0})^{\alpha_0} (\sum_{j_1=0}^{p^1} c_{a,j_1}^1 u^{j_1})^{\alpha_1} \cdots (\sum_{j_r=0}^{p^r} c_{a,j_r}^r u^{j_r})^{\alpha_r}.$$

By the multinomial theorem, we expand each factor $(\sum_{j_k=0}^{p^k} c_{a,j_k}^k u^{j_k})^{\alpha_k}$ in this expression into a polynomial $\sum_{j=0}^{\alpha_k p^k} \tilde{c}_{a,j}^k u^j$ in u of degree $\alpha_k p^k$. Since each multinomial coefficient is an integer, and these have p-adic norms bounded above by 1, by the multiplicative and non-Archimedean property of $\| \ \|_p$, and referring to the estimates of Lemma 7.1, we conclude that

$$\left\| \tilde{c}_{a,j}^k \right\|_p \leq \frac{1}{p^{(n-k)j}}.$$

It follows that

$$\prod_{k=0}^r (\sum_{j_k=0}^{\alpha_k p^k} \tilde{c}_{a,j_k}^k u^{j_k}) = \sum_{j=0}^{\sum_{k=0}^r \alpha_k p^k} \bar{c}_{a,j}^\alpha u^j,$$

where we may estimate the coefficient $\bar{c}_{a,j}^\alpha$ p-adically by

$$\left\| \bar{c}_{a,j}^\alpha \right\|_p \leq \frac{1}{p^{(n-r)j}}.$$

Therefore, if we choose and fix n to be any integer greater than the order r of the operator $f(x) = F(x, \delta x, \ldots, \delta^r x)$, the coefficients $\bar{c}_{a,j}^\alpha$ will all have uniformly bounded p-adic norm, and we have that

$$f(a + p^n u) = \sum_{\alpha=(\alpha_0,\ldots,\alpha_r)} a_\alpha \left(\sum_{j=0}^{\sum_{k=0}^r \alpha_k p^k} \bar{c}_{a,j}^\alpha u^j \right)$$

is a power series in u whose coefficients go to zero p-adically as $|\alpha| \to \infty$ because so do the a_αs. This finishes the proof. $\qquad \square$

8

Characteristic functions of discs in $\mathbb{Z}_p$: p-adic coordinates

It is now natural to attempt to find the analytic functions on $\mathbb{Z}_p$ that are δ-functions. The set of characteristic functions of discs in $\mathbb{Z}_p$ is a dense subset of $C(\mathbb{Z}_p, \mathbb{Q}_p)$ [36], so we investigate this problem for these characteristic functions. Thus, given the covering $\{j + p^n\mathbb{Z}_p\}_{j=0}^{p^n-1}$ of $\mathbb{Z}_p$ by discs of radius $1/p^n$, we ask if the characteristic function of any of these discs can be realized as an arithmetic differential operator. We shall initially concentrate most of our attention in the case $n = 1$. The explicit results we obtain in this case will guide the rest of our work.

Throughout this Chapter we use the coordinates of elements of $\mathbb{Z}_p$ that result when choosing the complete residue system $\{0, 1, \ldots, p-1\}$ in realizing their expansions (2.4). These are referred to as the standard p-adic coordinates.

8.1 Characteristic functions of discs of radii $1/p$

We consider first the case where $p = 2$. We ask if an operator of order 1 can be the characteristic function of a disc in the cover $\{j + 2\mathbb{Z}_p\}_{j=0}^{1}$ of $\mathbb{Z}_2$. Of course, it suffices to treat the question for the disc centered at the origin, so we ask if we can find coefficients $a_{m,n}$ so that the order 1 arithmetic operator

$$f(x) = \sum_{m,n\geq 0} a_{m,n} x^m (\delta x)^n \,,$$

can be the characteristic function of the disc $2\mathbb{Z}_2$.

We use the changes $x = 2u$ and $x = 1 + 2u$ on the discs $2\mathbb{Z}_2$ and $1 + 2\mathbb{Z}_2$, respectively. Then, for this to be the case, we must have the relations

$$
\begin{aligned}
f_0(u) &= \sum_{m,n\geq 0} a_{m,n}(2u)^m(u-2u^2)^n = 1\,, \\
f_1(u) &= \sum_{m,n\geq 0} a_{m,n}(1+2u)^m(-1)^n(u+2u^2)^n = 0\,,
\end{aligned}
\tag{8.1}
$$

as series in u. And this system admits 2-adic integer solutions for the a_{mn}s that go to zero as $m+n \to \infty$.

Indeed, a simple commutation argument yields the explicit expressions

$$
f_0(u) \;=\; a_{0,0} + \sum_{k=1}^{\infty} c_{0,k} u^k\,,
\tag{8.2}
$$

where

$$
\begin{aligned}
c_{0,k} \;=\; & \left(\sum_{n=[\frac{k}{2}]+}^{k} 2^{k-n} \sum_{m=0}^{k-n} a_{m,n}(-1)^{k-m-n} \binom{n}{k-m-n} \right. \\
& \left. + \sum_{n=0}^{[\frac{k-1}{2}]} 2^{k-n} \sum_{m=k-2n}^{k-n} a_{m,n}(-1)^{k-m-n} \binom{n}{k-m-n} \right)\,,
\end{aligned}
$$

while

$$
f_1(u) \;=\; \sum_{m=0}^{\infty} a_{m,0} + \sum_{k=1}^{\infty} c_{1,k} u^k\,,
\tag{8.3}
$$

where

$$
\begin{aligned}
c_{1,k} \;=\; & \left(\sum_{m=k}^{\infty} \sum_{n=0}^{k} a_{m,n}(-1)^n 2^{k-n} \binom{m+n}{k-n} + \right. \\
& \left. \sum_{m=0}^{k-1} \sum_{n=[\frac{k-m}{2}]+}^{k} a_{m,n}(-1)^n 2^{k-n} \binom{m+n}{k-n} \right)\,,
\end{aligned}
$$

respectively. Here, $[\,\cdot\,]$ and $[\,\cdot\,]^+$ are the greatest integer less or equal than and the smallest integer greater or equal than functions, respectively. Strassman's theorem [47] (see Theorem 4.9 in §4.1), (8.1) implies that all but one of the coefficients above vanish. We have:

Theorem 8.1   *The characteristic function of $2\mathbb{Z}_2$ is an arithmetic differential operator of order 1. The system of equations (8.1) can be solved for the coefficients $a_{m,n}$, with $a_{m,n}=0$ for all $m \geq 2$, and*

$$
\|a_{m,n}\|_2 \leq \frac{1}{2^n}\,.
$$

Proof. We proceed by induction on n. Since $a_{m,n} = 0$ for $m \geq 2$, the two equations in (8.1) suffice to solve for $a_{0,n}$ and $a_{1,n}$, respectively. Indeed, by (8.2), the coefficient of u^0 in $f_0(u)$ is $a_{0,0}$, so $a_{0,0} = 1$. By (8.3), the said coefficient in $f_1(u)$ is $\sum_{m=0}^{\infty} a_{m,0} = a_{0,0} + a_{1,0}$, and so $a_{1,0} = -1$. Similarly, the coefficient of u in $f_0(u)$ is $a_{0,1} + 2a_{1,0}$, so $a_{0,1} =$

2, and the said coefficient in the series $f_1(u)$ is given by $2\sum_{m=1}^{\infty} m a_{m,0} - \sum_{m=1}^{\infty} a_{m,1} - a_{0,1} = 2a_{1,0} - a_{1,1} - a_{0,1} = 0$, and that yields $a_{1,1} = -4$.

Assume that all the $a_{m,n}$s have been chosen for all pairs m, n with $n < k$, and satisfying the desired estimates for some $k \geq 1$. Since the coefficient $c_{0,k}$ of $f_0(u)$ must be zero, using its explicit form given by (8.2), we solve the resulting equation for $a_{0,k}$, and obtain a linear combination of the previously found coefficients. In this combination, $a_{m,n}$ appears multiplied by 2^{k-n} times some rational integer. Thus, $a_{0,k}$ is uniquely determined, and it satisfies the desired estimate.

Having found $a_{0,k}$, the coefficient $c_{1,k}$ of u^k in the series $f_1(u)$ given explicitly in (8.3), which must also be zero. allows us to solve for $a_{1,k}$ as a linear combination of the previously found coefficients, including $a_{0,k}$. In this combination, $a_{m,n}$ appears again multiplied by 2^{k-n} times some rational integer. Thus, $a_{1,k}$ is uniquely determined also, and it satisfies the desired estimate. This finishes the proof. $\square$

We now generalize the result above to discs of radius $1/p$ for an arbitrary prime p. Our argument extends the one just given for the prime 2, realizing the said characteristic function as an operator of the form $\sum_{n \geq 0} \sum_{m=0}^{p-1} a_{m,n} x^m (w_1 \delta(x) + \cdots + w_{p-1+d_p} \delta^{p-1+d_p}(x))^n$, where the w_js are suitably chosen coefficients, and where d_p is the degree of p, as defined below. A different generalization will be given in Chapter 9, where we will obtain a stronger result.

We begin our work here by associating to any prime p a canonical set of p-adic numbers. In order to do so, we define first the associated matrix A of p.

For $p = 2$, let A to be the number 1, a 1×1 matrix. For $p \geq 3$ and any integer $l \geq 0$, we let v_l be the vector

$$v_l = \begin{pmatrix} \delta^l(2) \\ \delta^l(3) \\ \vdots \\ \delta^l(p-1) \end{pmatrix}, \tag{8.4}$$

and define A to be the $(p-2) \times (p-2)$ matrix whose columns are the vectors $v_1, \ldots, v_{p-2}$:

$$A := \begin{pmatrix} \delta(2) & \delta^2(2) & \cdots & \delta^{p-2}(2) \\ \delta(3) & \delta^2(3) & \cdots & \delta^{p-2}(3) \\ \vdots & \vdots & \ddots & \vdots \\ \delta(p-1) & \delta^2(p-1) & \cdots & \delta^{p-2}(p-1) \end{pmatrix}. \tag{8.5}$$

We call A the matrix associated to p.

Definition 8.2   The Wronskian of the prime p is the determinant $\det A$ of its associated matrix A. The degree d_p of p is the p-adic order of its Wronskian. We say that p is a singular prime if its degree d_p is positive. Otherwise, p is said to be nonsingular. $\square$

Example 8.3   Generally speaking, it is a rather difficult task to compute the Wronskian of a prime p. For $p = 2$ it is 1. For $p = 3$, it is -2, and for $p = 5$, it is the rational integer

$$2^{23} \cdot 3^{19} \cdot 7^7 \cdot 11 \cdot 13^4 \cdot 17 \cdot 29 \cdot 61 \cdot 89 \cdot 1595154208691889882755666696509 \,.$$

Thus, the first 3 primes are all nonsingular.

In fact, among the first 100 primes, only $p = 29$ and $p = 311$ are singular. This assertion can be proven without having to find explicitly their Wronskian, merely proving that this Wronskian is divisible by p. Even then, the quick growth of the integers involved in the calculations makes the task computationally complex when the prime in question is moderately large. $\square$

Let p be a prime of order d_p. For any l in the range from 1 to $p - 2$, we define A_l to be the matrix obtained from the matrix A in (8.5) associated to p by replacing its l-th column by the vector v_{p-1+d_p} in (8.4). The following conclusion is recorded for use later on. It follows by Cramer's rule.

Lemma 8.4   *The vector*

$$\frac{1}{\det A}(-\det A_1, \ldots, -\det A_{p-2})$$

is the unique solution of the equation $Ax = -v_{p-1+d_p}$.

Definition 8.5   The ordered $(p - 1 + d_p)$-tuple

$$\frac{1}{\det A}(-\det A_1, \ldots, -\det A_{p-2}, 0, \ldots, 0, \det A)$$

is the set of Wronskian quotients associated to p. $\square$

Example 8.6   If p is nonsingular, the Wronskian quotients associated to p are all p-adic integers. And once again, these numbers are difficult to compute also. For $p = 3$, they are the integers $(1, 1)$. For $p = 5$, we use the value of the Wronskian given in Example 8.3, and obtain that the third Wronskian quotient associated to the prime 5 is given by

$$-\frac{\det A_3}{\det A} = \frac{3 \cdot 13 \cdot 31 \cdot 37 \cdot 251 \cdot 9324919439243 \cdot q}{159515420869188988827556696509},$$

where q is a prime whose decimal expansion contains 222 digits*.   □

Remark 8.7 The naive thought that the set of of Wronskian quotients associated to a prime p should all be p-adic integers fails for $p = 29$, the first of the singular primes.   □

Lemma 8.8 *Let d_p be the degree of the prime p, and let the $(p-1+d_p)$- tuple $(w_1, \ldots, w_{p-1+d_p})$ be the set of Wronskian quotients associated to it. Then the δ-function*

$$h(x) = w_1 \delta(x) + \cdots + w_{p-1+d_p} \delta^{p-1+d_p}(x)$$

vanishes at $x = a$, $0 \le a \le p - 1$, and if

$$h(a + pu) = \sum_{n=1}^{p^{p-1+d_p}} w_{a,j} u^j, \quad 0 \le a \le p-1,$$

then $\|w_{a,1}\|_p = p^{p-2+d_p}$, and $\left\|(w_{a,j}/w_{a,1}^j)\right\|_p \le 1/p^{j-1}$ for any j.

Proof. For a vector $w = (w_1, \ldots, w_{p-1+d_p})$ of constants yet to be determined, we consider the function

$$h(x) = w_1 \delta(x) + \cdots + w_{p-1+d_p} \delta^{p-1+d_p}(x).$$

Since regardless of what the positive integer k may be both 0 and 1 are in the kernel of $\delta^k(x)$, the set of conditions $h(a+pu)|_{u=0} = 0$, $0 \le a \le p-1$ yields the system of equations

$$\begin{pmatrix} \delta(2) & \delta^2(2) & \cdots & \delta^{p-1+d_p}(2) \\ \delta(3) & \delta^2(3) & \cdots & \delta^{p-1+d_p}(3) \\ \vdots & \vdots & \ddots & \vdots \\ \delta(p-1) & \delta^2(p-1) & \cdots & \delta^{p-1+d_p}(p-1) \end{pmatrix} \begin{pmatrix} w_1 \\ w_2 \\ \vdots \\ w_{p-1+d_p} \end{pmatrix} = 0.$$

Notice that the matrix of this system is simply the augmented matrix $(A \mid v_{p-1} \ldots v_{p-1+d_p})$. Since its row echelon form is given by the matrix

* $q = 469485807722695200929680463444552973880363492744895666211497702977996111969322231331735328135927753617578318483953518947685436992331399697089142870509915060400522731910953084863542351281032595851575511339337818 2840897402417.$

$(\mathbb{1} \mid A^{-1}v_{p-1} \ldots A^{-1}v_{p-1+d_p})$, we know that the solution space of the system above has $(d_p + 1)$-degrees of freedom. By Lemma 8.4, the set of Wronskian quotients $(w_1, \ldots, w_{p-1+d_p})$ associated to p is the unique solution of this system whose last $(d_p + 1)$-components are $(0, \ldots, 0, 1)$. We choose and fix this particular solution hereafter.

We pick a rational integer a such that $0 \leq a \leq p - 1$. Following the notation in Lemma 7.1, we let $c_{a,j}^k$ denote the coefficient of u^j in the polynomial $\delta^k(a + pu)$. By Remark 7.2, $c_{a,1}^k = p^{1-k} \prod_{l=0}^{k-1} (1 - p(\delta^l(a))^{p-1})$, and if $w_{a,j}$ denotes the coefficient of u^j in $h(a + pu)$, we have that

$$w_{a,1} = \sum_{k=1}^{p-1+d_p} \frac{w_k}{p^{k-1}} \prod_{l=0}^{k-1} (1 - p(\delta^l(a))^{p-1}).$$

Since $w_{p-1+d_p} = 1$, by the non-Archimedean property of $\| \ \|_p$, we see that $\|w_{a,1}\|_p \leq p^{p-2+d_p}$. In fact, we have equality. For otherwise, the expression above for $w_{a,1}$ multiplied by p^{p-2+d_p} would lead to the contradictory statement that $\|w_{p-1+d_p}\|_p < 1$. This uses the fact that if $d_p > 0$, the k-th Wronskian quotient associated to p is zero for ks such that $p - 1 \leq k \leq p - 2 + d_p$, and that all of these numbers times p^{d_p} are p-adic integers. Thus, $\|w_{a,1}\|_p = p^{p-2+d_p} = \left\| c_{a,1}^{p-1+d_p} \right\|_p$.

In general, we may write the $w_{a,j}$ in terms of the $c_{a,j}^k$s for arbitrary values of j as

$$w_{a,j} = w_1 c_{a,j}^1 + \cdots + w_{p-1+d_p} c_{a,j}^{p-1+d_p}.$$

If we now use Lemma 7.1 with $n = 1$ and the non-Archimedean property of $\| \ \|_p$, we obtain that

$$\left\| \frac{w_{a,j}}{w_{a,1}^j} \right\|_p = \left\| p^{(p-2+d_p)j} w_{a,j} \right\|_p \leq \frac{1}{p^{(p-2+d_p)j}} \max_{1 \leq k \leq p-1+d_p} \left\{ \left\| c_{a,j}^k \right\|_p \right\}$$

$$\leq \frac{1}{p^{j-1}}.$$

This completes the proof. $\square$

Remark 8.9   Given the Wronskian quotients $w_1, \ldots, w_{p-1+d_p}$ associated to p, and the coefficients $w_{a,j}$ of u^j in the polynomial function $h(a + pu)$ above, the scaling indicated in this result seems to be optimal if we are to make the most general statement. For instance, we could ask instead if the quotients $w_{a,j}/w_{a,1}$ are p-adic integers. This is true for the primes $p = 2$ and $p = 3$. It is false already for $p = 5$. $\square$

Remark 8.10   Notice that if p is a nonsingular prime, the arithmetic operator $h(x)$ in Lemma 8.8 is given by a polynomial in $\mathbb{Z}_p[x_0, \ldots, x_{p-1}]$ (see Definition 5.3), and has order $p-1$. Else, it is given by a polynomial in $\mathbb{Q}_p[x_0, \ldots, x_{p-1+d_p}]$. In both cases, the polynomials are independent of x_0.

Theorem 8.11   *Let d_p be the degree of p. Then, given a disc in $\mathbb{Z}_p$ of radius $1/p$, there exists a formal power series of the form*

$$\sum_{n \geq 0} \sum_{m=0}^{p-1} a_{m,n} x^m \left(w_1 \delta(x) + \cdots + w_{p-1+d_p} \delta^{p-1+d_p}(x) \right)^n ,$$

such that $a_{m,n} \to 0$ p-adically, which converges pointwise to its characteristic function. Here, $(w_1, \ldots, w_{p-1+d_p})$ are the Wronskian quotients associated to p.

In particular, when all of the Wronskian quotients of p are p-adic integers (if at all), we may conclude that the characteristic function of a disc of radius $1/p$ is an arithmetic differential operator of order $p-1+d_p$ of the form indicated in the Theorem. Thus, for nonsingular primes p the said characteristic functions are arithmetic differential operators of order $p-1$. For other primes, the assertion is only valid at the level of formal power series that converge pointwise.

Proof. It suffices to show that the characteristic function of $p\mathbb{Z}_p$ is a formal operator of order $p-1+d_p$ of the type indicated in the statement. The argument that we provide works for any prime p, and generalizes the one we have seen already in Theorem 8.1 for the case of the nonsingular prime $p = 2$.

Let $w_1, \ldots, w_{p-1+d_p}$ be the set of Wronskian quotients associated to the prime p, and consider the $(p - 1 + d_p)$-th arithmetic operator

$$h(x) = w_1 \delta(x) + \cdots + w_{p-1+d_p} \delta^{p-1+d_p}(x)$$

of Lemma 8.8. Then we know that for any a such that $0 \leq a < p$, we have that

$$h(a + pu) = \sum_{j=1}^{p^{p-1+d_p}} w_{a,j} u^j ,$$

where $\left\| w_{a,1} \right\|_p = p^{p-2+d_p}$ and $\left\| w_{a,j}/w_{a,1}^j \right\|_p \leq 1/p^{j-1}$. We proceed by induction to determine the coefficients $a_{m,n}$ in the operator

$$f(x) = \sum_{n \geq 0} \sum_{m=0}^{p-1} a_{m,n} x^m (h(x))^n$$

so that, if $f_a(u) = f(a + pu)$, we have that

$$
\begin{cases}
f_0(u) & = & 1, \\
f_1(u) & = & 0, \\
& \vdots & \\
f_{p-1}(u) & = & 0,
\end{cases}
\tag{8.6}
$$

and

$$\|a_{m,n}\|_p \leq \frac{1}{p^n}, \quad n = 0, 1, 2, \ldots \tag{8.7}$$

For the starting point of the induction, we choose the coefficients $a_{m,0}$, $0 \leq m < p$, such that

$$F_0(x) = \sum_{m=0}^{p-1} a_{m,0} x^m$$

solves (8.6) to order zero. This implies that $a_{0,0}$ is 1, and that the equation

$$
\begin{pmatrix}
1 & 1 & \cdots & 1 \\
2 & 2^2 & \cdots & 2^{p-1} \\
\vdots & \vdots & \ddots & \vdots \\
p-1 & (p-1)^2 & \cdots & (p-1)^{p-1}
\end{pmatrix}
\begin{pmatrix}
a_{1,0} \\
a_{2,0} \\
\vdots \\
a_{p-1,0}
\end{pmatrix}
=
\begin{pmatrix}
-1 \\
-1 \\
\vdots \\
-1
\end{pmatrix}
$$

holds. This Vandermonde system can be readily solved for the $a_{m,0}$s, with the solution being a vector of p-adic integers, and the estimates (8.7) holding for $n = 0$.

For the k-th step of the induction, we assume that we have found

$$F_{k-1}(x) = \sum_{n=0}^{k-1} \sum_{m=0}^{p-1} a_{m,n} x^m (h(x))^n$$

such that $F_a^{k-1}(u) = F_{k-1}(a + pu)$ solves system (8.6) to order $k - 1$, with the required estimates for the coefficients up to that point. Let us consider

$$F_k(x) = \sum_{n=0}^{k-1} \sum_{m=0}^{p-1} a_{m,n} x^m (h(x))^n + \sum_{m=0}^{p-1} a_{m,k} x^m (h(x))^k .$$

We determine the coefficients $a_{m,k}$ so that $F_a^k(u) = F_k(a + pu)$ solves (8.6) to order k.

The coefficient of u^k in $F_a^k(u)$ is determined by the $a_{m,n}$s with $n \le k$. In fact, by the multinomial expansion, this coefficient is given by

$$w_{a,1}^k \sum_{m=0}^{p-1} a_{m,k} a^m + \sum_{n=0}^{k-1} \sum_{m=0}^{p-1} a_{m,n} \sum_{r,\alpha} c_{n,\alpha,m,r,a,p} \prod_{l=1}^{p^{p-1+d_p}} w_{a,l}^{\alpha_l},$$

where

$$c_{n,\alpha,m,r,a,p} = \binom{n}{\alpha_1, \ldots, \alpha_{p^{p-1+d_p}}} \binom{m}{r} a^{m-r} p^r$$

and where the last sum in the second term above is over all nonnegative multi-indices α of weight n and indices r in the range $0 \le r \le m$, such that $\alpha_1 + 2\alpha_2 + \cdots + p^{p-1+d_p} \alpha_{p^{p-1+d_p}} + r = k$. Hence, the $F_a^k(u)$ solves (8.6) to order k if we have

$$\sum_{m=0}^{p-1} a_{m,k} a^m = - \sum_{n=0}^{k-1} \sum_{m=0}^{p-1} a_{m,n} \sum_{r,\alpha} c_{n,\alpha,m,r,a,p} \frac{\prod_{l=1}^{p^{p-1+d_p}} w_{a,l}^{\alpha_l}}{w_{a,1}^k} \quad (8.8)$$

for all as in the range $0 \le a \le p - 1$.

Notice that

$$\frac{\prod_{l=1}^{p^{p-1+d_p}} w_{a,l}^{\alpha_l}}{w_{a,1}^k} = \frac{1}{w_{a,1}^r} \prod_{l=2}^{p^{p-1+d_p}} \left(\frac{w_{a,l}}{w_{a,1}^l} \right)^{\alpha_l}.$$

By Lemma 8.8, it follows that the p-adic norm of this number is bounded by $\frac{1}{p^{k-n+(p-1+d_p)r}}$. Now by the induction hypothesis on the p-adic norm of the $a_{m,n}$s, we conclude that the right side of (8.8) has p-adic norm no larger than $1/p^k$.

The equation for $a_{0,k}$ is uncoupled from the remaining $a_{m,k}$s. Indeed, when $a = 0$ the only nonzero term in the left side of (8.8) is $a_{0,k}$. Thus, we have $a_{0,k} = g_0^k$, where g_0^k is a p-adic number with p-adic norm no larger than $1/p^k$. For the other coefficients, (8.8) yields the system

$$\begin{pmatrix} 1 & 1 & \cdots & 1 \\ 2 & 2^2 & \cdots & 2^{p-1} \\ \vdots & \vdots & \ddots & \vdots \\ p-1 & (p-1)^2 & \cdots & (p-1)^{p-1} \end{pmatrix} \begin{pmatrix} a_{1,k} \\ a_{2,k} \\ \vdots \\ a_{p-1,k} \end{pmatrix} = \begin{pmatrix} g_1^k \\ g_2^k \\ \vdots \\ g_{p-1}^k \end{pmatrix}$$

where the right side is a vector of p-adic numbers whose components all have p-adic norm no larger than $1/p^k$. This system can be solved with the solution having p-adic norm equal to the p-adic norm of the right side. This completes the last step in the induction, and the proof. $\square$

8.2   Characteristic functions of discs of radii $1/p^n$

The argument given above can be generalized in a rather naive manner for the study of the same problem for discs of radii $1/p^n$ as well. However, we should expect now a singular behaviour for $n > 1$. Indeed, if we think of p-adic numbers as analytic functions, as we have been doing so far, larger powers of p should exhibit a behaviour analogous to branching points. Our naive generalization shows that, and we obtain formal power series in $\mathbb{Q}_p[[x_0, \ldots, x_n]]$ which converge pointwise to the characteristic function of a disc of radius $1/p^n$, but the series are not necessarily restricted. As in the next Chapter we shall improve substantially upon our current result to this point, we merely outline the generalized argument, leaving several of the details to the interested reader.

We associate with p^n the $(p^n - 2) \times (p^n - 2)$ matrix given by

$$A_{p^n} := \begin{pmatrix} \delta(2) & \delta^2(2) & \cdots & \delta^{p^n-2}(2) \\ \delta(3) & \delta^2(3) & \cdots & \delta^{p^n-2}(3) \\ \vdots & \vdots & \ddots & \vdots \\ \delta(p^n - 1) & \delta^2(p^n - 1) & \cdots & \delta^{p^n-2}(p^n - 1) \end{pmatrix}. \tag{8.9}$$

Definition 8.12   We say that $\det A_{p^n}$ is the Wronskian of p^n, and call its p-adic order the degree of p^n, which we denote by d_{p^n}. $\square$

The integer $d_{p^n} > 0$ measures the singularity of p^n. By the computational complexity of the problem, it is a difficult task to find ps such that $d_{p^n} > 0$ for some $n > 1$.

We can define the ordered $(p^n - 1 + d_{p^n})$-tuple

$$\frac{1}{\det A_{p^n}}(-\det A_1, \ldots, -\det A_{p^n-2}, 0, \ldots, 0, \det A_{p^n})$$

as the Wronskian quotients associated to p^n. Here, A_j is the matrix obtained from A_{p^n} by replacing its j-th column by the vector whose k component is $\delta^{p^n-1+d_{p^n}}(2 + (k - 1))$, $1 \leq k \leq p^n - 2$. We then have:

Lemma 8.13   *Let d_{p^n} be the degree of p^n, and let $(w_1, \ldots, w_{p^n-1+d_{p^n}})$ be the set of Wronskian quotients associated to it. Then the δ-function*

$$h_n(x) = w_1 \delta(x) + \cdots + w_{p^n-1+d_{p^n}} \delta^{p^n-1+d_{p^n}}(x)$$

vanishes at $x = a$, $0 \leq a \leq p^n - 1$, and if

$$h(a + pu) = \sum_{n=1}^{p^{p^n-1+d_{p^n}}} w_{a,j} u^j, \quad 0 \leq a \leq p^n - 1,$$

then $\|w_{a,1}\|_p = p^{p^n-2+d_{p^n}}$, and $\left\|(w_{a,j}/w_{a,1}^j)\right\|_p \leq 1/p^{j-1}$ for any j. $\square$

We now consider n-tuples $\beta = (\beta_0, \ldots, \beta_{n-1})$ in $\mathbb{Z}^n$ such that $0 \leq \beta_0, \ldots, \beta_{n-1} \leq p - 1$. This set contains p^n elements. We provide it with the lexicographical order.

Lemma 8.14   *There are polynomials $p_0(x), \ldots, p_{n-1}(x)$ in $\mathbb{Z}[x]$ such that the $p^n \times p^n$ matrix $V = (v_{ij})$ whose entries are given by*

$$v_{ij} = (p_0(i))^{\beta_0}(p_1(\delta(i)))^{\beta_1} \cdots (p_{n-1}(\delta^{n-1}(i)))^{\beta_{n-1}}, \ 0 \leq i \leq p^n - 1,$$

where $(\beta_0, \ldots, \beta_{n-1})$ is the j-th element of the set $\{\beta = (\beta_0, \ldots, \beta_{n-1}) \in \mathbb{Z}^n : 0 \leq \beta_0, \ldots, \beta_{n-1} \leq p - 1\}$ in the lexicographical order, is invertible over $\mathbb{Z}_p$, that is to say, its determinant is an invertible element of $\mathbb{Z}_p$.

We skip the details of the proof. A more elaborate argument will be given in Lemma 9.4 below, where a closely related result is proven using the constant coordinates of the arithmetic differential theory. In that argument, we can see that the statement above can be proven by taking $p_j(x) = x$ for all j.

Based on this result, the generalized version of Theorem 8.11 for discs of arbitrary radii reads as follows.

Theorem 8.15   *Let p be a prime and $n \in \mathbb{N}$. Assume that d_{p^n} is the degree of p^n, and consider the disc $p^n\mathbb{Z}_p$ in $\mathbb{Z}_p$ of radius $1/p^n$. Let $h_n(x)$ and $p_0(x), \ldots, p_{n-1}(x)$ be the arithmetic differential operator and polynomials of Lemmas 8.13 and 8.14 above, respectively. Then, there exists a restricted power series of the form*

$$F(x_0, \ldots, x_n) = \sum_{m \geq 0} \sum_{0 \leq \beta_j < p} a_{\beta,n}(p_0(x_0))^{\beta_0} \cdots (p_{n-1}(x_{n-1}))^{\beta_{n-1}}(x_n)^m$$

in $\mathbb{Z}_p[[x_0, \ldots, x_n]]$ such that $F(x, \delta(x), \ldots, \delta^{n-1}(x), h_n(x))$ converges to the characteristic function of the said disc pointwise.

The argument parallels the one for $n = 1$ that proves Theorem 8.11. In the earlier case, the family of polynomials given by Lemma 8.14 consists of the single polynomial $p_0(x) = x$. In the general argument, the $p^n \times p^n$-matrix V given in Lemma 8.14 now plays the role of that of the Vandermonde matrix in the proof of Theorem 8.11.

Remark 8.16   The the representation of the characteristic function of any disc in $\mathbb{Z}_p$ as an arithmetic differential operator is not unique. $\square$

9

Characteristic functions of discs in $\mathbb{Z}_p$: harmonic arithmetic coordinates

Theorems 8.11 and 8.15 in the previous Chapter do not provide any indication of the optimal order required to realize characteristic functions of discs as arithmetic differential operators of the said order. This indication lacks even for nonsingular primes, where the order given by Theorem 8.11 is the lowest. In this Chapter we prove a result that is somewhat of a surprise, addressing this issue. It ties up the order of the operator that realizes the characteristic function of the disc with its level of analyticity (see Definition 4.4), and in that sense, it strengthens what Theorem 7.3 says about them.

No characteristic function of a disc of radius $1/p$ can be an operator of order zero, so both Theorems 8.1 & 8.11 yield the optimal result for the prime 2. However, what about the remaining primes?

Remark 9.1 No nontrivial arithmetic differential operator

$$f(x) = F(x, \delta x, \ldots, \delta^r x)$$

with $F(t_0, \ldots, t_r) \in \mathbb{Z}_p[t_0, \ldots, t_r]$ can equal a nontrivial locally constant function, no matter the order. For such a function would be a polynomial in x, of a certain degree, that cannot be zero. Since the degree k in u of $f_{a,n}(u) = f(a + p^n u)$ must be the same as the degree of $f(x)$ in x, and since for sufficiently large n, this function would have to be constant, by taking k analytic derivatives of $f_{a,n}$, and setting $u = 0$, we would see that $k = 0$. So $F(t_0, \ldots, t_r)$ must be the constant polynomial $\quad\square$

In the work carried out in the previous Chapter, we used the coordinate representation of the p-adic number that arise by considering the complete residue system $\{0, 1, \ldots, p-1\}$ as the coefficients in the expansion (2.4). In the context of arithmetic differential operators, we will

119

now show that it is far more natural to consider the coordinates that arise from the "arithmetic constants," the solutions of $\delta(a) = 0$. These are the $(p-1)$-roots of unity and zero, and they lead to the Teichmüller representation of Example 4.8. In a sense, these are analogous to the harmonic coordinates on a Riemannian manifold, although the analogy is quite incomplete in that we do not have a notion of Riemannian metric on $\mathbb{Z}_p$. However, the judicious use of these coordinates lead to the proof of the following result [14]:

Theorem 9.2   *Any analytic function $f : \mathbb{Z}_p \to \mathbb{Z}_p$ of level m is an arithmetic differential operator of order m.*

As it will turn out, the characteristic functions of discs in $\mathbb{Z}_p$ of radii $1/p^m$, which are analytic functions of level m, can be realized as arithmetic operators of the form $F(x, \delta x, \dots, \delta^m x)$, where $F(x_0, \dots, x_m)$ is a restricted power series in $\mathbb{Z}_p[[x_0, \dots, x_m]]$. This is a vast improvement on the representation that, for instance, Theorem 8.11 yields for arbitrary primes, even the nonsingular ones (cf. with the remark that follows the statement of the theorem, prior to its proof).

We have already mentioned that representation of a characteristic function of a disc in $\mathbb{Z}_p$ as an arithmetic differential operator is not unique. Of all of these representations, those given by restricted power series with coefficients in $\mathbb{Z}_p$ itself are to be preferred. They are the most natural examples from the point of view of number theoretic considerations.

9.1   A matrix associated to p^m

Let us consider the set of all p-adic integer roots of the function $x \mapsto \delta^m x$:

$$C_m := \{a \in \mathbb{Z}_p : \delta^m a = 0\} \subset \mathbb{Z}_p.$$

Since the m-th iterate of the operator δ is given by a polynomial in its argument of degree p^m with $\mathbb{Q}_p$-coefficients, C_m can have at most p^m elements. In fact, its cardinality is p^m as the number of solutions of $\delta^m a = 0$ is exactly p^m. For we have.

Lemma 9.3   *The composition*

$$C_m \hookrightarrow \mathbb{Z}_p \to \mathbb{Z}_p/p^m\mathbb{Z}_p$$

is bijective.

Proof. We prove the assertion by induction on m. For $m = 0$, we have that $C_0 = \{0\}$, and the result is clear. We assume now that the statement is true for $m - 1 > 0$, and prove it for m.

Given $a \in C_{m-1}$, we consider the polynomial $t^p - t + pa \in \mathbb{Z}_p[t]$. By Hensel's lemma, it has p distinct roots that we denote by $a_1, \ldots, a_p \in \mathbb{Z}_p$. Notice that we have $\delta a_j = a$ for all js, and since $\delta^{m-1}a = 0$, it follows that $\delta^m a_j = 0$.

Observe that if $a, a' \in C_{m-1}$ and we have that

$$a_j \equiv a'_{j'} \mod p^m \tag{9.1}$$

for some j, j', then $a = a'$ and $j = j'$. Indeed, if (9.1) holds then $a \equiv a'$ mod p^{m-1}, and by the induction hypothesis, $a = a'$, and hence $j = j'$ as well.

We thus have that each of the p^{m-1} element of C_{m-1} yields p distinct elements of C_m, so C_m contains a set of p^m elements, and by the observation above, this set injects into $\mathbb{Z}_p/p^m\mathbb{Z}_p$. As C_m has at most p^m elements, this forces the map $C_m \to \mathbb{Z}_p/p^m\mathbb{Z}_p$ to be bijective. $\square$

We use Lemma 9.3 to write

$$C_m = \{a_0, a_1, \ldots, a_{p^m-1}\}, \tag{9.2}$$

where $a_\alpha \equiv \alpha \mod p^m$ for all αs in the set

$$I = \{0, \ldots, p^m - 1\}. \tag{9.3}$$

On the other hand, we fix an ordering of the set

$$I' = \{\beta = (\beta_0, \ldots, \beta_{m-1}) \in \mathbb{Z}^m : 0 \leq \beta_0, \ldots, \beta_{m-1} \leq p - 1\}, \tag{9.4}$$

and consider the $p^m \times p^m$-matrix

$$W = (w_{\alpha\beta})_{\alpha \in I, \beta \in I'}, \tag{9.5}$$

whose entries are given by

$$w_{\alpha\beta} := (a_\alpha)^{\beta_0}(\delta a_\alpha)^{\beta_1} \cdots (\delta^{m-1}a_\alpha)^{\beta_{m-1}} \in \mathbb{Z}_p,$$

where we have used the convention that $a^0 = 1$ for all $a \in \mathbb{Z}_p$. Up to a permutation of its columns, the matrix W is intrinsically associated to the number p^m.

Lemma 9.4 *The determinant of the matrix W in (9.5) is invertible in $\mathbb{Z}_p$.*

Proof. We use the reduction mod p mapping

$$\mathbb{Z}_p \;\to\; \mathbb{F}_p := \mathbb{Z}_p/p\mathbb{Z}_p$$
$$a \;\mapsto\; \bar{a}.$$

Now let us notice that if $a' \equiv a \bmod p^m$ then $\delta a' \equiv \delta a \bmod p^{m-1}$. For we have that $a' = a + p^m u$ with $u \in \mathbb{Z}_p$. So

$$
\delta a' = \frac{a' - (a')^p}{p} \;=\; \frac{a + p^j u - (a + p^j u)^p}{p}
$$
$$
= \;\frac{a - a^p}{p} + p^{m-1}u' = \delta a + p^{m-1}u',
$$

for some $u' \in \mathbb{Z}_p$, and so $\delta a' \equiv \delta a \bmod p^{m-1}$. We may iterate this argument to conclude that $\delta^i a' \equiv \delta^i a \bmod p^{m-i}$ for any $i \le m$, and so the function

$$\mathbb{Z}_p \;\to\; \mathbb{F}_p^m$$
$$a \;\mapsto\; (\bar{a}, \overline{\delta a}, \ldots, \overline{\delta^{m-1}a})$$

induces a bijection between $\mathbb{Z}_p/p^m\mathbb{Z}_p$ and $\mathbb{F}_p^m$. Now for any element $\gamma = (\gamma_0, \ldots, \gamma_{m-1})$ in $\mathbb{F}_p^m$ and any $\beta \in I'$, we set

$$
v_{\gamma\beta} = \gamma_0^{\beta_0} \gamma_1^{\beta_1} \cdots \gamma_{m-1}^{\beta_{m-1}}.
$$

Notice that for any $a_\alpha \in C_m$ we have that $\delta^i a_\alpha \equiv \delta^i \alpha \bmod p$ if $i \le m-1$. Therefore, the desired result here will follow by Lemma 9.3 if we merely show that $\det(v_{\gamma\beta}) \ne 0 \in \mathbb{F}_p$.

Let us assume that $\det(v_{\gamma\beta}) = 0$. This means that there exist constants $\lambda_{\beta_0 \ldots \beta_{m-1}} \in \mathbb{F}_p$ for $(\beta_0, \ldots, \beta_{m-1}) \in I'$, not all zero, such that

$$
\sum_{\beta_0=0}^{p-1} \cdots \sum_{\beta_{m-1}=0}^{p-1} \lambda_{\beta_0 \ldots \beta_{m-1}} \gamma_0^{\beta_0} \gamma_1^{\beta_1} \cdots \gamma_{m-1}^{\beta_{m-1}} = 0
$$

for all $\gamma \in \mathbb{F}_p^m$. We may then proceed by induction on m to prove that these relations imply that that all the λs vanish, which is a contradiction. This finishes the proof. $\qquad\square$

9.2 Analytic functions and arithmetic differential operators

We now carry out the proof of Theorem 9.2 by proving the following stronger result. In what follows, I and I' are the sets of indices (9.3) and (9.4), respectively.

Theorem 9.5 *Let $f : \mathbb{Z}_p \to \mathbb{Z}_p$ be an analytic function of level m. Then there exists a unique restricted power series $F \in \mathbb{Z}_p[[x_0, x_1, \ldots, x_m]]$ with the following properties:*

1. $F(x_0, x_1, \ldots, x_m) = \sum_{n \geq 0} \sum_{\beta \in I'} a_{\beta,n} x_0^{\beta_0} x_1^{\beta_1} \cdots x_{m-1}^{\beta_{m-1}} x_m^n.$
2. $f(a) = F(a, \delta a, \ldots, \delta^m a), \ a \in \mathbb{Z}_p.$

Proof. We start by proving the existence of the power series F. Notice that if

$$g : \mathbb{Z}_p \to \mathbb{Z}_p$$

is any arithmetic differential operator of order m, and $a \in \mathbb{Z}_p$, then

$$h(x) := g(x + a)$$

is also an arithmetic differential operator of order m; cf. Lemma 5.4. Thus, without losing generality, we may assume that the function f in the statement of the Theorem is zero on all discs of radius $1/p^m$ except for $p^m \mathbb{Z}_p$. Without losing generality also, we may additionally assume that there exists an $l \geq 0$ such that $f(p^m u) = u^l$ for all $u \in \mathbb{Z}_p$.

We recall the set $C_m = \{a_0, \ldots, a_{p^m-1}\}$ in (9.2). The family of discs $\{a + p^m \mathbb{Z}_p\}_{a \in C_m}$ forms a covering of $\mathbb{Z}_p$. Let us notice that $C_m \ni a_0 = 0$, and this is the center of the one disc where f is nonzero.

By Lemma 7.1, for $a \in C_m$ and $0 \leq k \leq m$, we have that $\delta^k(a + p^m u) = \sum_{j=0}^{p^m} c_{a,j}^k u^j$, with $c_{a,0}^k = \delta^k a$, and

$$\left\| c_{a,0}^k \right\|_p \leq 1, \quad \left\| c_{a,1}^k \right\|_p = \frac{1}{p^{m-k}}, \quad \left\| c_{a,j}^k \right\|_p \leq \frac{1}{p^{(m-k+1)j-1}}, \ 2 \leq j \leq p^k.$$
$$(9.6)$$

We may view $\delta^k(a + p^m u)$ as an element in the ring of polynomials $\mathbb{Z}_p[u]$. Since $\delta^m a = 0$, we have that $c_{a,0}^m = 0$.

We proceed to determine by induction the coefficients $a_{\beta,n}$ in the series F appearing in the statement so that

$$\left\| a_{\beta,n} \right\|_p \leq \min \left\{ 1, \frac{1}{p^{n-l}} \right\}, \quad n \geq 0, \ \beta \in I',$$
$$(9.7)$$

and so that, if $F_a(u) = F(a + p^m u)$ for $a \in C_m$, then we have that

$$\begin{cases} F_a(u) &= u^l \quad \text{if} \quad a = 0, \\ F_a(u) &= 0 \quad \text{if} \quad a \neq 0. \end{cases}$$
$$(9.8)$$

Here we view (9.8) as equalities of functions of $u \in \mathbb{Z}_p$. However, since each $F_a(u)$ is defined by a restricted power series in $\mathbb{Z}_p[[u]]$, it is enough to check (9.8) as equalities in the ring of formal power series $\mathbb{Z}_p[[u]]$.

We consider the polynomials $F_a^k(u) \in \mathbb{Z}_p[u]$ defined by

$$F_a^k(u) := \sum_{n=0}^{k} \sum_{\beta \in I'} a_{\beta,n} \left(\sum_{j=0}^{p^0} c_{a,j}^0 u^j \right)^{\beta_0} \cdots \left(\sum_{j=0}^{p^{m-1}} c_{a,j}^{m-1} u^j \right)^{\beta_{m-1}} \left(\sum_{j=1}^{p^m} c_{a,j}^m u^j \right)^n$$

so that $F_a^k(u)$ converge u-adically to F_a in $\mathbb{Z}_p[[u]]$. We find the $a_{\beta,n}$s inductively so they satisfy the estimate (9.7), and such that the following congruences hold in the ring $\mathbb{Z}_p[u]$:

$$\begin{cases} F_a^k(u) \equiv u^l \mod u^{k+1} & \text{if } a = 0, \\ F_a^k(u) \equiv 0 \mod u^{k+1} & \text{if } a \neq 0. \end{cases} \qquad (9.9)$$

In what follows we denote by δ_{ij} the Kronecker symbol.

For the starting point of the induction, we choose the coefficients $a_{\beta,0}$, $\beta \in I'$, such that (9.9) and (9.7) hold. This can be achieved by solving the system of equations

$$\sum_{\beta \in I'} w_{\alpha\beta} a_{\beta,0} = \delta_{l0}\delta_{\alpha 0}, \quad \alpha \in I,$$

where $W = (w_{\alpha\beta})$ is the matrix (9.5). By Lemma 9.4, this system can be readily solved for the $a_{\beta,0}$s, with the solution being a vector of p-adic integers.

For the k-th step of the induction, let us notice that for $a = a_\alpha$, the coefficient of u^k in $F_a^k(u)$ is given by

$$(c_{a,1}^m)^k \sum_{\beta \in I'} w_{\alpha\beta} a_{\beta,k} + \sum_{n=0}^{k-1} \sum_{\beta \in I'} a_{\beta,n} b_{\beta,n,k}, \qquad (9.10)$$

where $b_{\beta,n,k}$ is the coefficient of u^k in

$$\left(\sum_{j=0}^{p^0} c_{a,j}^0 u^j \right)^{\beta_0} \cdots \left(\sum_{j=0}^{p^{m-1}} c_{a,j}^{m-1} u^j \right)^{\beta_{m-1}} \left(\sum_{j=1}^{p^m} c_{a,j}^m u^j \right)^n.$$

Thus, $b_{\beta,n,k}$ is a $\mathbb{Z}$-linear combination of products of the form

$$\left(\prod_{i=1}^{\beta_0} c_{a,j_{0i}}^0 \right) \cdots \left(\prod_{i=1}^{\beta_{m-1}} c_{a,j_{m-1,i}}^{m-1} \right) \left(\prod_{i=1}^{n} c_{a,j_{mi}}^m \right),$$

with

$$\sum_{r=0}^{m-1} \sum_{i=1}^{\beta_r} j_{ri} + \sum_{i=1}^{n} j_{mi} = k.$$

We may assume that there are integers s_r such that $j_{ri} \geq 1$ for $i \leq s_r$ and $j_{ri} = 0$ for $i > s_r$. So we have

$$s_r \leq \sum_{i=1}^{s_r} j_{ri}, \quad \sum_{r=0}^{m-1} \sum_{i=1}^{s_r} j_{ri} + \sum_{i=1}^{n} j_{mi} = k. \tag{9.11}$$

By (9.6) and the induction hypothesis,

$$\|a_{\beta,n} b_{\beta,n,k}\|_p \leq \min\left\{1, \frac{1}{p^{n-l+\sigma}}\right\}. \tag{9.12}$$

where

$$\sigma = \sum_{r=0}^{m-1} [(m-r+1)(\sum_{i=1}^{s_r} j_{ri}) - s_r] + \sum_{i=1}^{n} j_{mi} - n.$$

Now, by (9.11) we have that

$$\begin{aligned}
\sigma &\geq \textstyle\sum_{r=0}^{m-1}[2(\sum_{i=1}^{s_r} j_{ri}) - s_r] + \sum_{i=1}^{n} j_{mi} - n \\
&\geq \textstyle\sum_{r=0}^{m-1} \sum_{i=1}^{s_r} j_{ri} + \sum_{i=1}^{n} j_{mi} - n \\
&= k - n.
\end{aligned}$$

Hence

$$\|a_{\beta,n} b_{\beta,n,k}\|_p \leq \min\left\{1, \frac{1}{p^{k-l}}\right\}. \tag{9.13}$$

Now, by the induction hypothesis also, we can ensure that $F_a^k(u)$ satisfies (9.9) if we have that

$$\sum_{\beta \in I'} w_{\alpha\beta} a_{\beta,k} = (c_{a_\alpha,1}^m)^{-k} \left(\delta_{kl}\delta_{\alpha 0} - \sum_{n=0}^{k-1} \sum_{\beta \in I'} a_{\beta,n} b_{\beta,n,k} \right), \quad \alpha \in I. \tag{9.14}$$

By (9.6) and (9.13), the p-adic norm of the right hand side of (9.14) is bounded above by $\min\{1, 1/p^{k-l}\}$. Again, by Lemma 9.5, we can solve the system (9.14) for the $a_{\beta,k}$s, with the solution satisfying the estimates (9.7). This completes the induction, and hence the existence part of the Theorem.

In order to prove the uniqueness, we need to show that if a restricted power series F satisfies conditions (1) and (2) in the Theorem for $f = 0$, then $a_{\beta,n} = 0$ for all $\beta \in I'$, $n \geq 0$. This can derived by an induction on n, in view of the equalities

$$\sum_{\beta \in I'} w_{\alpha\beta} a_{\beta,k} = -(c_{a_\alpha,1}^m)^{-k} \left(\sum_{n=0}^{k-1} \sum_{\beta \in I'} a_{\beta,n} b_{\beta,n,k} \right), \quad \alpha \in I.$$

If $n = 0$, this expression says that $\sum_{\beta \in I'} w_{\alpha\beta} a_{\beta,0} = 0$, $\alpha \in I$, and since the matrix $(w_{\alpha\beta})$ is invertible, we have that $a_{\beta,0} = 0$ for all $\beta \in I'$. For larger values of n, the expression then says that $\sum_{\beta \in I'} w_{\alpha\beta} a_{\beta,k} = 0$ also for $\alpha \in I$, since according to it, these values are determined by the $a_{\beta k'}$s with $0 \leq k' < k$, that are zero. So $a_{\beta,k} = 0$ for all $\beta \in I'$. This completes the proof. $\qquad\square$

10

Some differences between arithmetic differential operators over $\mathbb{Z}_p$ and $\widehat{\mathbb{Z}}_p^{ur}$

In this Chapter we explore some differences between arithmetic differential operators over the ring $\widehat{\mathbb{Z}}_p^{ur}$ and arithmetic differential operators over the coarser ring $\mathbb{Z}_p$. In particular, we discuss the naive analogue of Theorem 9.2 over $\widehat{\mathbb{Z}}_p^{ur}$, which we show to be false.

Indeed, in the context of the general theory of [6, 8] discussed in §6, where the role of the ring $\mathbb{Z}_p$ is played by $\widehat{\mathbb{Z}}_p^{ur}$ and the operator δ is associated to the lift of Frobenius (6.2) of Theorem 6.9, we now have the following:

Theorem 10.1 *Let*

$$F(x) = \sum a_\alpha x_0^{\alpha_0} \cdots x_r^{\alpha_r}$$

be a restricted power series in $\widehat{\mathbb{Z}}_p^{ur}[[x]]$. Assume that the mapping

$$x \mapsto F(x, \delta x, \ldots, \delta^r x)$$

is constant on a disk of some radius in $\widehat{\mathbb{Z}}_p^{ur}$. Then $F(x)$ itself is a constant in $\widehat{\mathbb{Z}}_p^{ur}$.

The proof of this result requires to use the fact that $\overline{\mathbb{F}}_p = \widehat{\mathbb{Z}}_p^{ur}/p\widehat{\mathbb{Z}}_p^{ur}$ is algebraically closed. This explains, in part, our earlier assertion in Chapter 6 that the differences between the theories of arithmetic differential operators over the rings $\mathbb{Z}_p$ and $\widehat{\mathbb{Z}}_p^{ur}$ are analogous to the differences between number theoretic statements about finite fields and algebraic geometric statements over their algebraic closures.

We begin the proof of Theorem 10.1 by recalling a key result in [11]. For convenience, we denote by R the ground ring, as before. If X is a smooth scheme over R of finite type, and $\{U_i\}$ is a covering of X by affine open sets, we recall that the p-adic completions of the schemes

Spec $\mathcal{O}(J^n(U_i))$ glue together to produce the formal scheme $J_p^n(X)$, the p-jet space of X of order n. If we glue the p-adic completions of Spec $\mathcal{O}(J^\infty(U_i))$ instead, we obtain the infinite p-jet space of X. Since the sheaf of rings $\mathcal{O}(J_p^n(X))$ are topologically generated by $\mathcal{O}(J_p^{n-1}(X))$ and $\delta\mathcal{O}(J_p^{n-1}(X))$, we obtain a projective system of formal schemes (6.5), where $J_p^0(X) = X^{\widehat{(p)}}$ is the p-adic completion of X, and $J_p^\infty(X)$ is the p-adic completion of the inverse limit of the $J_p^n(X)$s. By taking the reduction mod p, we obtain a projective system of $k = R/pR$-schemes

$$\cdots \to J_0^r(X) \to J_0^{r-1}(X) \to \cdots \to J_0^1(X) \to J_0^0(X) = X_0^{\widehat{(p)}}.$$

By the universality property of the p-jets, there exists a natural lifting mapping

$$\nabla : X(R) \to J_p^\infty(X)(R).$$

In the case where X is the affine line over R, this mapping is just

$$R \overset{\nabla}{\to} R^{\mathbb{N}}$$
$$x \mapsto \nabla x = (x, \delta x, \delta^2 x, \ldots).$$

By composition with the reduction mod p mapping, we obtain the mapping

$$R \overset{\nabla_0}{\to} k^{\mathbb{N}}$$
$$x \mapsto \nabla_0 x = (\overline{x}, \overline{\delta x}, \overline{\delta^2 x}, \ldots),$$

where we denote by $\overline{x}$ the reduction $x \bmod p$ of an element x in R.

We shall need the following only in a particular case, but state it in general. A Witt vector over a commutative ring A is a sequence $x = (x_0, x_1, x_2, \ldots)$ of elements of A. The Witt polynomials are defined by $W_n(x) = \sum_{i=0}^n p^i x_i^{p^{n-i}}$, $n = 0, 1, \ldots$. Then there exists a unique ring structure on the set of Witt vectors over A such that:

1. The addition and product operations are given by universal polynomials with integral coefficients, and
2. Every Witt polynomial is a homomorphism from the ring of Witt vectors over A into A.

For instance, we have that

$$(x_0, x_1, \ldots) + (y_0, y_1, \ldots) = \left(x_0 + y_0, x_1 + y_1 + \frac{x_0^p + y_0^p - (x_0 + y_0)^p}{p}, \ldots \right),$$
$$(x_0, x_1, \ldots) \times (y_0, y_1, \ldots) = \left(x_0 y_0, x_0^p y_1 + y_0^p x_1 + p x_1 y_1, \ldots \right).$$

Remark 10.2   A p-derivation δ_p over a ring A is a theory of length two of the ring of Witt vectors of A, that is to say, a theory that involves the ring $W_2(A)$ of the first two components of a Witt vector of A in a manner compatible with the ring structure they inherit from that of the Witt vectors. Or said differently, if A is a ring and B is an algebra over A, a p-derivation $\delta_p : A \to B$ associated to the homomorphism $\phi_p : A \to B$ is such that the map $A \ni a \mapsto (a, \delta_p(a)) \in W_2(B)$ is a ring homomorphism, where in the right side, the first component a stands for its image in the B algebra over A. $\square$

Example 10.3   1. The Witt ring of any commutative ring A in which p is invertible is isomorphic to $A^{\mathbb{N}}$. For the Witt polynomials produce a homomorphism from the ring of Witt vectors to $A^{\mathbb{N}}$, and if p is invertible, the said homomorphism is an isomorphism.

2. Consider the field $\mathbb{F}_p = \mathbb{Z}_p/p\mathbb{Z}_p$. Then its Witt ring is $\mathbb{Z}_p$. On the other hand, we have that the algebraic closure $\overline{\mathbb{F}}_p$ is equal to $\overline{\mathbb{F}}_p = \mathbb{Z}_p^{ur}/p\mathbb{Z}_p^{ur}$, and every $\overline{x} \in \overline{\mathbb{F}}_p$ has a unique Teichmüller representative $x \in \mathbb{Z}_p^{ur}$ (see Example 4.8), which is a root of unity and projects to $\overline{x}$ in $\mathbb{Z}_p^{ur}/p\mathbb{Z}_p^{ur}$. It follows that the Witt ring of $\overline{\mathbb{F}}_p$ is $\widehat{\mathbb{Z}}_p^{ur}$. $\square$

Assume now that R is a commutative local ring with maximal ideal pR such that the residue field $k = R/pR$ is of characteristic p. Of course, this will be the case if $R = \widehat{\mathbb{Z}}_p^{ur}$. Then the ring $W(k)$ of Witt vectors of k is $k^{\mathbb{N}}$. If the ring R is Hausdorff and complete for the topology of ideals defined by $pR \supset p^2 R \supset \cdots$, and if $k = R/pR$ is a perfect ring of characteristic p, there exists one and only one multiplicative system of representatives $\psi : k \to R$ [44]. These are multiplicative mappings such that $\pi\psi = \mathbb{1}$, and such that

$$k^{\mathbb{N}} \xrightarrow{\theta} R$$
$$a = (\alpha_0, \alpha_1, \dots) \quad \mapsto \quad \theta(a) = \sum_{j=0}^{\infty} \psi(a_j^{p-j})p^j$$

is a ring isomorphism.

Lemma 10.4   (Lemma 2.6, [11]) *The composition mapping*

$$k^{\mathbb{N}} \xrightarrow{\theta} R \xrightarrow{\nabla_0} k^{\mathbb{N}}$$

is a bijection. In fact, there exist integer coefficients universal polynomials P_n in n-variables, $n \geq 2$, such that

$$\nabla_0\theta(\alpha_0, \dots, \alpha_n, \dots) = (\alpha_0, \alpha_1, \alpha_2 + P_2(\alpha_0, \alpha_1), \dots, \alpha_n + P_n(\alpha_0, \dots, \alpha_{n-1}), \dots).$$

We now have the following.

Lemma 10.5   *Let $F(x) = \sum a_\alpha x_0^{\alpha_0} \cdots x_r^{\alpha_r} \in R[[x]]$ be a restricted power series such that $F(x, \ldots, \delta^r x) = 0$ for all $x \in R$. Let $\overline{F}$ denote the reduction mod p of F. Then $\overline{F} = 0$ in the polynomial ring $k[w_0, \ldots, w_r]$.*

Proof. Since $F(x, \ldots, \delta^r x) = 0$ for all $x \in R$, we have that

$$\overline{F}(\overline{x}, \overline{\delta x}, \ldots, \overline{\delta^r x}) = 0$$

for all $x \in R$. By Lemma 10.4 above, any vector $w = (w_0, w_1, \ldots, w_r) \in k^{r+1}$ is the projection onto the first $(r+1)$-components of a vector $\nabla_0(\theta(a)) = (\overline{a}, \overline{\delta a}, \ldots, \overline{\delta^r a}, \ldots)$. Thus, $\overline{F}(x) = \overline{F}(w_0, w_1, \ldots, w_r) = 0$ for any $w \in k^{r+1}$, as desired.   □

Notice that both Lemmas 10.4 and 10.5 hold in general, without requiring the quotient field $k = R/pR$ to be algebraically closed. This hypothesis is needed in order to complete the proof of Theorem 10.1, as we now see.

Proof of Theorem 10.1. By subtracting the constant, we can assume that $F(x, \delta x, \ldots, \delta^r x) = 0$, and so in order to prove the result, it will suffice to prove that $p|F$, for by iteration we will then able to conclude that F must be identically zero. By Lemma 10.5, we have that

$$\overline{F}(w_0, \ldots, w_r) = 0$$

for any $(w_0, \ldots, w_r) \in k^{r+1}$. Thus, $\overline{F}$ is in the zero ideal of the polynomial ring $k[w]$. Since the residue field $k = \widehat{\mathbb{Z}}_p^{ur}/p\widehat{\mathbb{Z}}_p^{ur}$ is algebraically closed, we may apply Hilbert's Nullstellensatz to conclude that $\overline{F} = 0$. This finishes the proof.   □

The most general assertion that can be made along these lines is the following:

Theorem 10.6   *Let X is a smooth scheme over $R = \widehat{\mathbb{Z}}_p^{ur}$. If f is an element in the ring $\mathcal{O}(J_p^r(X))$ of global functions on the p-jet space of order r such that the induced map $X(R) \to R$ is 0, then f itself is 0.*

The proof of this generalized version can be given in a manner parallel to the one above. We merely need to replace the use of Lemma 10.4 as stated here by its most general version in [11].

Remark 10.7   We may elaborate further on our results over the ring $\mathbb{Z}_p$, and compare them to Mahler's theorem 3.2 about the structure of continuous $\mathbb{Z}_p$-valued functions on $\mathbb{Z}_p$.

Let $\mathbb{Z}_0^\infty$ be the set of all nonnegative integral vectors $\alpha = (\alpha_0, \alpha_1, \alpha_2, \ldots)$

with finite support, that is to say, vectors $\alpha = (\alpha_0, \alpha_1, \alpha_2, \dots)$ in $\mathbb{Z}^\infty$ such that $\alpha_j \geq 0$ for all j, and $\alpha_j = 0$ for j sufficiently large. If $\alpha \in \mathbb{Z}_0^\infty$, the weight $|\alpha| = \sum_{i \geq 0} \alpha_i$ is well-defined. Given a sequence of variables $x_0, x_1, x_2, \dots$ and an $\alpha \in \mathbb{Z}_0^\infty$, we set x^α for $x_0^{\alpha_0} x_1^{\alpha_1} x_2^{\alpha_2} \dots$. In this case, we say that a power series

$$F(x_0, x_1, x_2, \dots) = \sum_{\alpha \in \mathbb{Z}_0^\infty} a_\alpha x^\alpha, \ a_\alpha \in \mathbb{Z}_p,$$

is *restricted* if $\lim_{|\alpha| \to \infty} a_\alpha = 0$.

Mahler's theorem 3.2 is equivalent to the following, as proven, for instance, in [18]:

Theorem 10.8   *Let $f : \mathbb{Z}_p \to \mathbb{Z}_p$ be a continuous function. Then there exists a restricted power series $F(x_0, x_1, x_2, \dots)$ in the variables $x_0, x_1, x_2, \dots$, with $\mathbb{Z}_p$-coefficients, such that*

$$f(a) = F(a, \delta a, \delta^2 a, \dots)$$

for all a in $\mathbb{Z}_p$.   □

Our Theorems 7.3 and 9.2 imply that the series F in Theorem 10.8 can be chosen to depend on finitely many variables if, and only if, f is analytic.

If in Definition 4.4 we replace $\mathbb{Z}_p$ by $\widehat{\mathbb{Z}}_p^{ur}$, we would then obtain the notion of an analytic function of level m over $\widehat{\mathbb{Z}}_p^{ur}$.

Theorem 10.9   *Let*

$$F(x) = \sum a_\alpha x_0^{\alpha_0} \cdots x_r^{\alpha_r}, \ r \geq 1,$$

be a restricted power series in $\widehat{\mathbb{Z}}_p^{ur}[[x]]$ such that the mapping

$$x \mapsto F(x, \delta x, \dots, \delta^r x)$$

is analytic. Then $F(x)$ must be a constant function. On the other hand, if $F(x)$ is an analytic function of level m on $\widehat{\mathbb{Z}}_p^{ur}$, it cannot be an arithmetic differential operator unless $m = 0$.

Thus, over $\widehat{\mathbb{Z}}_p^{ur}$, an arithmetic differential operator that is not of order zero is never analytic, and an analytic function that is not of level zero is never an arithmetic differential operator.

Proof. The first statement amounts to the uniqueness of the representation of an analytic function on $\widehat{\mathbb{Z}}_p^{ur}$, uniqueness that was proven combining Lemma 10.5 and the Nullstellensatz in the proof of Theorem 10.1, cf. with Remark 8.16. The last assertion then follows. $\qquad\square$

Remark 10.10   In Example 5.5 we exhibited the Legendre symbol $a \mapsto \left(\frac{a}{p}\right)$ over $\mathbb{Z}_p^\times$ as a differential operator of order 1. The right hand side of (5.7) makes sense for $a \in \widehat{\mathbb{Z}}_p^{ur}$ provided that δa is now defined by (6.1), where ϕ is the lift of Frobenius. The extended function is no longer locally constant, and, of course, it cannot be equal to the characteristic function of any disc. $\qquad\square$

Example 10.11   Example 5.5 has an analogue for affine elliptic curves. We discuss it briefly.

Let X be the locus of $v^2 = u^3 + au + b$ in the affine plane $\operatorname{Spec} \mathbb{Z}_p[u, v]$ over $\mathbb{Z}_p$, where $4v^3 + 27w^2$ is invertible, and view X as embedded in 3-space via the map $(v, w) \mapsto (v, w, (4v^3 + 27w^2)^{-1})$. Let $N(p, a, b)$ be the number of $\mathbb{F}_p$-points of X, that is to say, the number of points of the reduction mod p of the said curve. These can be expressed in terms of the traces of Frobenius $a_p(a, b)$, which are given as the coefficient of x^{p-1} in $(x^3 + ax + b)^{\frac{p-1}{2}}$.

The number of solutions $v \in \mathbb{F}_p$ to the equation $v^2 = w$ is equal to $1 + \left(\frac{w}{p}\right)$. Therefore, counting the point at ∞, we must have that

$$N(p, a, b) = 1 + \sum_{u \in \mathbb{F}_p} \left(1 + \left(\frac{u^3 + au + b}{p}\right)\right) = p + 1 + \sum_{u \in \mathbb{F}_p} \left(\frac{u^3 + au + b}{p}\right).$$

We could then try to use (5.7) to obtain an explicit expression for $N(p, a, b)$ as an arithmetic differential operator in a and b, but that will not work because we would obtain a series whose terms involve the denominator $(u^3 + au + b)^{np}$, and for each a and b, there could be a $u \in \mathbb{F}_p$ that annihilates it.

The identity $\left(\frac{a}{p}\right) \equiv a^{\frac{p-1}{2}} \bmod p$ is analogous to $N(p, a, b) \equiv -a_p(a, b) \bmod p$. The function $a_p(a, b)$ can can be represented [8, 10] as a quotient of arithmetic differential operators of order 2 defined on

$$X(\mathbb{Z}_p) = \{(a, b) \in \mathbb{Z}_p \times \mathbb{Z}_p : 4a^3 + 27b^2 \in \mathbb{Z}_p^\times\},$$

of certain restricted power series in $a, b, \delta a, \delta b, \delta^2 a, \delta^2 b$ and $(4a^3 + 27b^2)^{-1}$, though these series are not as explicit as that given by (5.7) for the Legendre symbol. We sketch the argument.

Let us recall that the $R[\phi_p]$-module of characters of an elliptic curve is generated by an operator f of order two or order one, depending upon the type of curve. Let $\omega = du/v$ be the standard invariant differential. Then we have a local expression

$$f(P) = \Lambda_f(p^{-1} \cdot \log_\omega P), \ P \in X(pR)$$

about the origin for a unique operator $\Lambda_f = \lambda_2 \phi_p^2 + \lambda_1 \phi_p + \lambda_0 \in R[\phi_p]$, the Picard-Fuchs operator associated to the character generator f. It can be proved [8, 10] that, depending upon the type of elliptic curve, we have that either $\Lambda_f = \phi_p^2 - a_p\phi_p + p$ or $\Lambda_f = \phi_p - \beta$, thus identifying fully the coefficients of Λ_f as an element of $R[\phi_p]$. On the other hand [10, Theorem 6.6], the Picard-Fuchs operator is proven to be of the form $\Lambda_f = \phi_p^2 - (f^2/f^1)\phi_p + h$ for certain arithmetic operators f^1, f^2 given by restricted power series in $a, b, \delta a, \delta b, \delta^2 a, \delta^2 b$ and $(4a^3 + 27b^2)^{-1}$. The desired assertion about $N(p, a, b)$ follows by a comparison of coefficients. $\square$

Theorems 9.2 and 10.1 show that in going from $\mathbb{Z}_p$ to $\widehat{\mathbb{Z}}_p^{ur}$ we loose the property of expressing the characteristic function of a disc as an arithmetic differential operator. Since the maximal unramified extension $\mathbb{Q}_p^{ur}$ of $\mathbb{Q}_p$ is obtained by adjoining to $\mathbb{Q}_p$ all the roots of unity of order relatively prime to p, and the set of unramified integers $\mathbb{Z}_p^{ur}$ is just $\mathbb{Z}_p^{ur} = \{x \in \mathbb{Q}_p^{ur} : \|x\|_p \leq 1\}$, it is natural to ask at what point we loose this property. We finish our work by proving that the said property is strictly a p-adic one.

Let $\zeta = \zeta_n$ be any root of unity of order n with n relatively prime to p, and let $\mathbb{Q}_p(\zeta)$ be the unramified extension of $\mathbb{Q}_p$ obtained by adjoining ζ. This is a vector space over $\mathbb{Q}_p$ whose dimension is the degree of ζ_n, which if ζ_n is a primitive root coincides with n. In what follows we do not loose generality if we make this assumption. We denote by $\mathbb{Z}_p(\zeta)$ the ring of integers of the extension, and let

$$\phi_{p,n} : \mathbb{Z}_p(\zeta_n) \to \mathbb{Z}_p(\zeta_n)$$

be the homomorphism associated to the δ_p operator on $\mathbb{Z}_p(\zeta)$.

Lemma 10.12   *The homomorphism* $\phi_{p,n}$ *is not an analytic function.*

Proof. Suppose that $\phi_{p,n}(x)$ is equal to a restricted power series on some disc of radius $1/p^N$. Without loss of generality, we may assume that the disc is centered at the origin in $\mathbb{Z}_p(\zeta_n)$. Since

$$\phi_{p,n}(x + y) = \phi_{p,n}(x) + \phi_{p,n}(y)$$

we may take (standard) derivatives with respect to y and evaluate the result at $y = 0$ to conclude that

$$\frac{d}{dx}\phi_{p,n}(x) = \frac{d}{dx}\phi_{p,n}(0),$$

and so the derivative is constant. Hence the power series giving $\phi_{p,n}(x)$ must contain linear or constant terms only. But since $\phi_{p,n}$ is a homomorphism that extends the lift of Frobenius, this implies that the power series and so $\phi_{p,n}$ itself must be the identity. This is a contradiction.

For if the identity were a lift of Frobenius on $\mathbb{Z}_p(\zeta_n)$, then it will induce the identity as a Frobenius mapping on the quotient $\mathbb{Z}_p(\zeta_n)/p\mathbb{Z}_p(\zeta_n)$. But the latter is a finite field, and the Frobenius mapping on a finite field is the identity if, and only if, the field is $\mathbb{F}_p$. □

The result above leads us to think that Theorem 9.2 fails to hold as soon as an unramified root of unity is adjoined to $\mathbb{Z}_p$. If we are to prove that result in this context, we can no longer appeal to the use of Hilbert's Nullstellensatz as in the proof of Therem 10.1.

Analytic functions in $\mathbb{Q}_p(\zeta_n)$ have unique power series expansion representations [43]. We may use this fact in proving the following result, the details of which we leave to the interested reader.

Theorem 10.13 *On $\mathbb{Z}_p(\xi_n)$, no arithmetic differential operator that is not of order zero is analytic, and no analytic function that is not of level zero is an arithmetic differential operator.*

References

[1] M.F. Atiyah & I.G. McDonald, *Introduction to commutative algebra.* Addison-Wesley Publishing Co., Reading, Mass.-London-Don Mills, Ont. 1969, ix+128 pp.

[2] G. Bachman, *Introduction to p-adic numbers and valuation theory*, Academic Press, New York-London 1964, ix+173 pp.

[3] D. Bertrand, *Sous groupes à un paramètre p-adiques de variétés de groupes*, Invent. Math. 40, 1977, pp. 171–193.

[4] J. Borger, *The basic geometry of Witt vectors*, arXiv:math/0801.1691.

[5] A. Buium, *Arithmetic analogues of derivations*, J. Algebra, 198, 1997, pp. 290–299.

[6] A. Buium, *Arithmetic Differential Equations*, Mathematical Surveys and Monographs, 118. American Mathematical Society, Providence, RI, 2005, xxxii+310 pp.

[7] A. Buium, *Differential Algebra and Diophantine Geometry*, Actualités Mathématiques. Hermann, Paris, 1994, x+182 pp.

[8] A. Buium, *Differential characters of Abelian varieties over p-adic fields*, Invent. Math. 122, 1995, pp. 309–340.

[9] A. Buium, *Differential modular forms*, J. Reine Angew. Math. 520, 2000, pp. 95–167.

[10] A. Buium, *Geometry of Fermat adeles*, Trans. Amer. Math. Soc. 357, 2005, pp. 901–964.

[11] A. Buium, *Geometry of p-jets*, Duke Math. J. 82, 1996, pp. 349–367.

[12] A. Buium, *Intersections in jet spaces and a conjecture of S. Lang*, Ann. of Math., 136, 1992, pp. 557–567.

[13] A. Buium & S.R. Simanca, *Arithmetic differential equations in several variables*, Ann. Inst. Fourier. Grenoble 59, 2009, pp. 2685–2708.

[14] A. Buium, C.C. Ralph & S.R. Simanca, *Arithmetic differential operators on $\mathbb{Z}_p$*, J. Number Theory 131, 2011, pp. 96–105.

[15] A. Buium & S.R. Simanca, *Arithmetic Partial Differential Equations, I*, Adv. Math. 225, 2010, pp. 689–793.

[16] A. Buium & S.R. Simanca, *Arithmetic partial differential equations, II: modular curves*, Adv. Math. 225, 2010, pp. 1308–1340.

[17] A. Buium & S.R. Simanca, *Arithmetic Laplacians*, Adv. Math. 220, 2009, pp. 246–277.

[18] P.-J. Cahen & J.-L. Chabert, *Integer Valued Polynomials*, Mathematical Surveys and Monographs, 48. American Mathematical Society, Providence, RI, 1997, xx+322 pp.

[19] P. Cassidy, *Differential algebraic groups*, Amer. J. Math. 94, 1972, pp. 891–954.

[20] C. Chevalley, *Class Field Theory*, Nagoya University, Nagoya, 1954, ii+104 pp.

[21] C. Chevalley, *La théorie du symbole de restes normiques*, J. reine angew. Math., 169, 1933, pp. 140–157.

[22] P. Deligne, *La conjecture de Weil*. I, Publ. Math. IHES, 43, 1974, pp. 273–307.

[23] B. Dwork, *On the rationality of the zeta function of an algebraic variety*, Amer. J. Math. 82, 1960, pp. 631–648.

[24] B. Dwork, *On the zeta function of a hypersurface*, Publ. Math. I.H.E.S. 12, 1962, pp. 5–68.

[25] B. Dwork, G. Gerotto & F.J. Sullivan, *An Introduction to G-functions*, Annals of Mathematical Studies 133, Princeton University Press, Princeton, New Jersey, 1994.

[26] R. Hartshorne, *Algebraic Geometry*, Graduate Texts in Mathematics, 52. Springer-Verlag, New York-Heidelberg, 2000, xvi+496 pp.

[27] H. Hasse, *Über die Darstellbarkeit von Zahlen durch quadratische Formen im Körper der rationalen Zahlen*, J. reine angew. Math., 152, 1923, pp. 129–148.

[28] K. Hensel, *Theorie der Algebraischen Zhalen*, Leipzig 1908.

[29] N. Koblitz, *p-adic numbers, p-adic analysis, and zeta-functions*. Second edition. Graduate Texts in Mathematics, 58. Springer-Verlag, New York, 1984, pp. xii+150.

[30] E.R. Kolchin, *Differential Algebra and Algebraic Groups*, Pure and Applied Mathematics, 54. Academic Press, New York-London, 1973, xviii+446 pp.

[31] I. Kra & S.R. Simanca, *On Circulant Matrices*, preprint 2011.

[32] T. Kubota & H.W. Leopoldt, *Eine p-adische theorie der zeta-werte I*, J. Reine und eigew. Math. 214/215, 1964, pp. 328–339.

[33] B. Mazur & A. Wiles, Class fields of abelian extensions of $\mathbb{Q}$, Invent. Math. 76, 1984, pp. 179-330,

[34] S. Lang, Cyclotomic Fields I and II. Combined second edition. With an appendix by Karl Rubin. Graduate Texts in Mathematics, 121. Springer-Verlag, New York, 1990, xviii+433 pp.

[35] K. Mahler, *Ein Beweis der Transzendenz der p-adische Exponenzialfunkzion*, J. reine angew. Math. 169, 1932, pp. 61–66.

[36] K. Mahler, *Introduction to p-adic numbers and their functions*, Cambridge Tracts in Mathematics. No. 64. Cambridge University Press, London-New York, 1973, ix+91 pp.

[37] Yu.I. Manin, *Algebraic curves over fields with differentiation*, Izv. Akad. Nauk SSSR, Ser. Mat. 22, 1958, pp. 737–756 = AMS translations Series 2, 37, 1964, pp. 59–78.

[38] Yu.I. Manin, *Periods of cusp forms and p-adic Hecke series*, Math. Sbornik., 92, 1973, pp. 378–401.

[39] A. Ostrowski, *Über einige Lösungen der Funktionalgleichung* $\varphi(x)\varphi(y) = \varphi(xy)$, Acta Math. 41, 1918, pp. 271–284.

[40] R. Palais, *Foundations of Global Non-Linear Analysis*, W. A. Benjamin, Inc., New York-Amsterdam 1968, vii+131 pp.

[41] J.F. Ritt, *Differential Algebra*, American Mathematical Society Colloquium Publications, XXXIII, American Mathematical Society, New York, N. Y., 1950, viii+184 pp.

[42] J-P. Serre, *A Course in Arithmetic*, Graduate Texts in Mathematics, 7. Springer-Verlag, New York-Heidelberg, 1973, viii+115 pp.

[43] J-P. Serre, *Lie algebras and Lie groups*, Benjamin, New York, 1965.

[44] J-P. Serre, *Local fields*, Graduate Texts in Mathematics, 67. Springer-Verlag, New York-Heidelberg, 1979, viii+241 pp.

[45] S.R. Simanca, *A mixed boundary value problem for the Laplacian*. Illinois J. Math. 32, 1988, pp. 98–114.

[46] S.R. Simanca, *Pseudo-differential operators*. 236, Pitman Research Notes in Mathematics, Longman Scientific & Technical, Harlow, 1990, 123 pp.

[47] R. Strassman, *Über den Wertevorrak von Potenzreihen im Gebiet der p-adischen Zahlen*, J. Reine Angew. Math. 159, 1928, pp. 13–28.

[48] A. Weil, *Number of solutions of equations in finite fields*, Bull. Amer. Math. Soc. 55, 1949, pp. 497–508.

Index

138